Praise for *Dog, Inc.*

"John Woestendiek's outstanding look at dog cloning explores what goes down when science, personal loss, and financial opportunism collide." —*Parade*

"Faded beauty queens, alien encounters, and conspiracy theories—these are a few of the things that Pulitzer Prize–wining investigative reporter John Woestendiek ran up against during his tour of the pet-cloning industry." —*The Daily*

"John Woestendiek has written a book about dog cloning that is clearly informed by a love of the animals but focuses on the people who exploit and manipulate them. . . . The inside story behind the costly quest to clone dogs reveals at least as much about human nature as it does about copying man's best friend." —*Psychology Today*

"Adorable little abominations of nature . . . They can clone Spot, but he's never the same dog." —*New York Post*

"The inside story behind the costly quest to clone dogs reveals at least as much about human nature as it does about copying man's best friend."

—*Alan Boyle,* MSNBC.com

"*Dog, Inc.* merits a close look, even if you're not in the market for a clone. As Woestendiek demonstrates, what was once science fiction is now our shared reality."

—*Richmond Times-Dispatch*

"Investigative reporter Woestendiek weaves together bizarrely interesting tales of rich pet owners, Korean and American scientists, ethics, and a petting zoo full of loved animals (including dogs, cats, and a Brahman bull). . . . From explaining the X-inactivation that foiled the results of the first cloned cat to relaying the story of Booger, a stray dog that learned to provide service to his injured mistress, Woestendiek educates as he entertains." —*Publishers Weekly*

"Woestendiek adroitly juxtaposes the inherent seriousness of the animal-human connection with the inanity of people who fork over big bucks for pet funerals, taxidermy, mummification and freeze-drying." —*Kirkus Reviews*

"Woestendiek expertly weaves multiple story lines, deftly creating a cloak that fairly covers the entire story. And he has pulled off a difficult feat, writing a science book that's highly entertaining." —*The Philadelphia Inquirer*

"A fascinating, thoroughly researched book that provides a close look at the controversial subject of cloning dogs." —*Modern Dog* magazine

"[*Dog, Inc.* is] a valuable contribution illuminating the hubris and futility of trying to replicate dead pets (or people) that will appeal to dog lovers and those interested in cloning and science." —*Library Journal*

Dog, Inc.

Dog, Inc.

How a Collection of Visionaries, Rebels, Eccentrics, and Their Pets Launched the Commercial Dog Cloning Industry

■ ■ ■

John Woestendiek

AVERY
a member of Penguin Group (USA) Inc.
New York

Published by the Penguin Group
Penguin Group (USA) Inc., 375 Hudson Street, New York, New York 10014, USA • Penguin Group (Canada), 90 Eglinton Avenue East, Suite 700, Toronto, Ontario M4P 2Y3, Canada (a division of Pearson Penguin Canada Inc.) • Penguin Books Ltd, 80 Strand, London WC2R 0RL, England • Penguin Ireland, 25 St Stephen's Green, Dublin 2, Ireland (a division of Penguin Books Ltd) • Penguin Group (Australia), 250 Camberwell Road, Camberwell, Victoria 3124, Australia (a division of Pearson Australia Group Pty Ltd) • Penguin Books India Pvt Ltd, 11 Community Centre, Panchsheel Park, New Delhi–110 017, India • Penguin Group (NZ), 67 Apollo Drive, Rosedale, North Shore 0632, New Zealand (a division of Pearson New Zealand Ltd) • Penguin Books (South Africa) (Pty) Ltd, 24 Sturdee Avenue, Rosebank, Johannesburg 2196, South Africa

Penguin Books Ltd, Registered Offices: 80 Strand, London WC2R 0RL, England

First trade paperback edition 2012

Published simultaneously in Canada

The Library of Congress catalogued the hardcover edition as follows:

Woestendiek, John.
Dog, Inc. : the uncanny inside story of cloning man's best friend / John Woestendiek.
p. cm.
ISBN 978-1-58333-391-4
1. Dogs—Cloning. 2. Human-animal relationships. I. Title.
QH442.2.W64 2010
636.7'0821—dc22
2010023697

ISBN 978-1-58333-464-5 (paperback edition)

Printed in the United States of America
1 3 5 7 9 10 8 6 4 2

Book design by Meighan Cavanaugh

To my mother,

my father,

my son,

and my dog

Contents

Prologue 1

1. Animal Control 9
2. The Missy Mission 20
3. Booger 27
4. Trakr 36
5. Sperling's Dogma 44
6. The Bullheaded Cowboy 56
7. The Raël World 70
8. "I Love Perfect Things" 80
9. Positively Cloneworthy 93
10. The Copied Cat 104
11. The Forever Pet 114
12. Good-bye, Dolly 125

13. Korea Joins the Race 140

14. Cats, Clones, Clowns, and Con Men 152

15. Second Chance 166

16. Snuppy 174

17. Booger at Rest 184

18. Dr. Hwang's Downfall 192

19. Missy Accomplished 205

20. The Turnspit Dog 214

21. Booger, Times Five 229

22. The Story of a Mormon Beauty Queen 237

23. Bringing Home the Clones 248

24. Trakr, Times Five 263

25. The End of the Dog Fight 273

Epilogue 285

Acknowledgments 299

Index 301

Prologue

■ ■ ■

> You might as well know it all right now. Lassie will not be meetin' you after school anymore.
>
> —Lassie Come Home *(film)*

HOUSTON, TEXAS
1967

It didn't matter if she was in a movie or a TV show; if the obstacles were raging river, forest fire, mountain range, or angry bear: Lassie always came home.

One day Tippy didn't.

On my birthday, in 1958, Tippy was my gift, a collie named for the white spot at the end of his tail—not a mirror image of the famous (to baby boomers, anyway) TV dog, but close enough for a five-year-old.

Unlike Lassie—the collie whose courage and loyalty were reincarnated in seven movies before starting a twenty-year run as an American TV show—Tippy never saved anyone stuck in a well. But, when he wasn't roaming the neighborhood, he did help my brother and me through childhood, frequent relocation, and our parents' divorce. When Dad left, Tippy stayed. When President Kennedy was shot and killed, Tippy was there to lean and cry on. An assassinated president may have served as my first long-distance lesson in mortality. But the demise of Tippy, my first dog, a few years later, brought death home.

That's one of the things dogs do for us. Often, as children, we learn through them that there's a limit to life—in their case, a very short one. What we're often slower to learn is how to accept that and fully celebrate them during the ten or fifteen years they're around. They come and go with class, while we—sometimes visibly, sometimes invisibly—fall apart at the seams.

When Tippy died, there were no such things as dog-loss support groups or dog bereavement counseling; people would have laughed at the very idea. There were few books on how to cope with your dog's death, few alternatives for disposing of the corpse, no websites on which to pay tribute to a dog that had passed. There were no agencies to guide one through the grief, and few companies, at the time, seeking to exploit it.

In the 1960s, when your dog died, you shed your tears in private and moved on, or at least pretended to.

Tippy lived mostly outside—that's what dogs did back then—first on Long Island, where he loved to trot two blocks down the hill and swim in Huntington Bay. When we moved to Houston, and in and out of three different neighborhoods, he continued his habit of wandering, perhaps because of an adventurous spirit, perhaps because he was never neutered, perhaps, in retrospect, because we weren't models of responsible pet ownership. Once he was found at a luxury hotel, four blocks away, where, apparently pining for the beach, he'd gone to frolic in the fountain.

He had a tricolored coat so thick it could muffle cries, so lush you could lose your fist in it, and he never objected to it being tugged. By hanging on to a handful of his fur, my little brother learned to walk. Tippy was, even with his occasional absences, patient and dependable. When I tripped up, he was there, with the kind of steady grace few humans master. He was my friend when, recently moved to a new neighborhood, I had none. He served as anchor, and lifeline. But he liked his alone time, too. One day he went off for some and didn't come back.

When we got word that someone had found a dead dog and buried it in the woods alongside the creek—and that, yes, it could have been a collie—I shifted into what I later learned was called denial. It couldn't be my dog. My dog always came home. Intent on proving wrong what everyone suspected, or perhaps seeking what I later learned was called closure, I grabbed a shovel and, along with my little brother, walked to the creek to see if the dog under the dirt was mine.

Seeing a mound of fresh soil, I started digging until, with a muffled thud, I hit fur. Tentatively scraping the dirt aside, I saw the familiar white and tan, tinged with black and shades of gray. Just a quick glance at a small patch of it, and I knew it was Tippy. I don't know if I covered him back up. I don't know if I cried. I don't think I experienced closure—just frustration and rage. I don't remember any adults telling me he was going to doggie heaven, or a place called "Rainbow Bridge" where dogs play happily in a meadow, awaiting the arrival of their masters before they proceed together into heaven.

At thirteen, I probably wouldn't have bought it if they did.

Not counting goldfish, hamsters, lightning bugs in jars, dyed Easter chicks, and captive toads, it was my first up-close experience with death, and it was in a whole different league than those earlier losses. They were backyard livestock, suburban curiosities, childhood science experiments. Tippy was a dog, my dog, at my side from not long after my first sounding out "See Spot run" all the way into the agony of adolescence. We'd become kindred spirits, in some ways reflections of each other, and, though I didn't know it then, what lay there in that shallow grave was, pretty much, what was left of my childhood.

Just a dog? Hardly. "Just a dog" is something the dog-less say, like, "You can always get another one." It takes a deep bond—a decade or more worth of shared experiences, love invested and exponentially returned—and diminishes it to the status of a broken toaster.

That you're not as likely to hear those words today as forty years ago is just one indication of what dogs have come to mean to us. To their owners, at least, every dog is exceptional, irreplaceable. If dogs were just dogs, their owners wouldn't be spending their life savings—and sometimes more—on dialysis, pacemakers, and other sophisticated medical procedures once restricted to humans, to keep their companions alive a few more years. If dogs were just dogs, industries wouldn't have arisen to memorialize, bury, cremate, and otherwise send them off into the great beyond. If dogs were just dogs, mankind probably would not have crossed the line from cloning for science and agriculture into the far murkier sector of cloning for love and money.

Over the years, losing a dog, given the increasingly close relationship we have with the species, has only gotten harder. The more we emotionally invest in dogs, the more we humanize them, the more we come to depend on each other, the more difficult still their deaths are to bear. They become part of the rhythm of our lives—the feeding, the walking, the quiet cuddle time. They, like us, are creatures of habit, and together we develop a cozy, synchronized daily routine. They tap into what's still pure and joyful and innocent inside us, no matter how deeply buried it may be. They keep the child in us alive.

We come to see them as extensions of ourselves, rightly or wrongly attributing to them qualities, emotions, and characteristics the pets may or may not have, but we feel they do. We often choose dogs who remind us of ourselves. And a good argument can be made that we—dog and guardian—come to resemble each other even more as we share a life together, forming a bond so resolute it becomes separable only by death.

When death approaches, we tend to fight tooth and nail. We, as a species, don't generally want to leave. So we assume dogs don't either. When illness and old age afflict them, we fight back—ostensibly on behalf of our dog, but, in truth, at least partly for selfish reasons.

Because when our dog dies, a part of us does, too—not just in a poetic way, not just metaphorically. A dog's death is a reminder of our own mortality, but it's also the passing of that innocent piece of us, that nurturing piece of us, that purposeful piece of us that we saw reflected, daily, in their soulful eyes. Oftentimes, the void they once filled reopens.

Even before cloning entered the picture and opened yet another door for the bereaved, or soon to be bereaved, pet owner, there was a tendency among many to replace dogs that had passed on with look-alikes—either the same breed or one chosen because of its resemblance to the departed—as if we were trying to get the same dog back. I'll admit to doing it twice.

We hate letting go, especially—as much as we talk about

"closure"—letting go entirely. And who's to say we shouldn't, or, for that matter, what forms our fight to preserve a beloved pet should take? We disagree on where the lines are, on what's reasonable, on what's "appropriate" and what's going overboard—when it comes to both prolonging a pet's life and preserving its memory. There is little consensus, and even where there is, it's irrelevant: to many a dog owner, what "everybody thinks" doesn't matter when their dog is in the clutches of death.

Given the evolution of the dog-human bond, and the far swifter evolution of biotechnology, the times could not have been riper for pet cloning to make its debut. As the twenty-first century approached, our love of dogs, the lengths to which we would go for them, and cloning technology were all on upward trajectories, destined to intersect, it seemed, in the years to follow. When they did, man, for the first time, would be able to re-create a loved one. Pet owners could look at death as something other than final. They could, in a fashion, hang on to their dog a little longer. They could refill the void and, perhaps, through a puppified version of their old dog, regain their own vitality.

In reality, pet-cloning clients aren't getting the same dog back, any more than all the dogs that have come in and out of my life and yours are all waiting for us patiently at "Rainbow Bridge." I don't believe Tippy is there. He'd have wandered off by now, just as I, though never totally letting go of my memories of him, have moved on to other dogs.

First, there was a Weimaraner I never fully connected with, though she did see fit to give birth to eleven pups in my bed. Then came a border collie who could not be trusted; a succession of communal college dogs; a golden retriever who either ran away or was stolen; a stray taken in off the streets; Auggie, my first wife's dog, who, not long after being put down, was followed by Hobo, another scruffy-coated black dog, who was followed by Fancy, another scruffy-coated black dog.

Fine dogs all, but none, to me, were the soul mate that Tippy had been, at least not until forty years later, when I—post-divorce and probably, in part, to fill a void—adopted a shelter mutt named Ace. He's a mix of Rottweiler, Akita, chow chow, and pit bull—four breeds of sometimes ill repute that merged into a gentle, 130-pound, curly-tailed giant, calm enough to serve as a therapy dog, sensitive enough to read my moods. One look into his big brown eyes and I'm at peace with the world. Sorrows aren't so sad. Disasters aren't so disastrous. Evils disappear.

I'm five again.

1.

Animal Control

■ ■ ■

It is not the strongest of the species that survives, nor the most intelligent, but the one most responsive to change.

—*Clarence Darrow*

SEOUL, SOUTH KOREA

November 2008

The agitated American was back.

She'd stood before the same ticket agents at the United Airlines counter in Seoul-Incheon International Airport the day before, and the one before that—pleading in tears one moment, loudly threatening lawsuits the next. She and her five nearly identical puppies needed to get home to California, and putting them in the jet's cargo area—as the airline was insisting its rules required—was, to her, out of the question.

Even after she presented them with some dubious "official" certificates stating the pups, despite their tender age, were service dogs, the airline officials held firm. She could carry one in her lap. The other four, they insisted, would have to travel as cargo.

"But I have three handicaps," Bernann McKinney countered, big blue eyes staring out from under blond bangs. "I should be allowed to take at least three dogs, one for each."

For hours a day, over a week's time, the argument continued—irresistible force versus immovable object, further complicated by language problems. Airline employees spoke only broken English; McKinney, despite having left her hometown in the mountains of North Carolina ten years earlier, still spoke in a syrupy southern drawl.

The puppies were too fragile and too valuable to travel in the cargo hold, she tried to explain, nicely at first. Federal law, she insisted a little more strongly, allows service dogs to travel in the cabin. She tried seeking pity: her money was running out, and her heart medication, too. She tried making a scene: she was, she argued, being kept in Korea against her will. She threatened to call the CEO of the airline, the police, and the U.S embassy. Nothing worked.

Each day's failure to book passage meant another $150 in taxi fare for McKinney, and having to find another dog-friendly hotel—a rare commodity in Seoul. She'd already been kicked out of four of them. One, she said, relegated her and the dogs to the boiler room. She had planned to spend six days in Seoul. Yet it had stretched to two weeks, with each day being a repeat of the previous one, like the movie *Groundhog Day*, but with five yapping, pooping puppies—all, like the passing days, nearly mirror images of each other.

Every morning, she checked out of a hotel and arrived at the airport with them—five dogs, wearing five service-dog vests, in five individual carrying bags. First, she sought travelers who might be willing to stretch the truth and carry a dog on board under the guise of being handicapped. She offered them free airfare to California and back in exchange for the favor. But even among those who spoke English and heard her out, she found no takers. That left her no recourse but to go back to the ticket counter, and, in what was almost a carbon copy of the previous day's exasperated exchange, plead her case again.

"You don't understand," she said finally. "These dogs are clones."

When that didn't work either, she gathered her mismatched luggage and her five genetically identical puppies—the world's first commercially produced canine clones—went outside, and hailed a cab. She'd try again tomorrow.

LOS ANGELES

June 2008

It could have been a scene from one of those old Publishers Clearing House commercials, where Ed McMahon and camera crew converge on a home, ring the doorbell, and inform the startled occupants that they have won an amount beyond their wildest dreams.

While it wasn't Ed at the door, and while the prize wasn't an oversized check, what transpired in Los Angeles on June 30, 2008—and was duly videotaped by the man behind it—was enough to make police-officer-turned-soap-opera-actor James Symington, once he was back inside his house, cry real tears.

A month earlier, Symington had entered an essay contest. He wrote an homage to his retired police dog, Trakr, a German shepherd who, though still alive, was ailing. In the essay, he recounted his dog's many achievements in the Halifax, Nova Scotia, police department—from locating stolen property to finding missing people. He addressed the special bond they shared. And he revisited the day in September 2001 when he, a friend, and Trakr—the dog retired by then—raced to New York City to help in any way they could, spending two days tracking through the rubble of the World Trade Center and locating, Symington says, the last surviving victim of the 9/11 attack.

Two years later, Symington left Halifax for Hollywood to pursue his dream of becoming an actor. Trakr went along, though his health was declining. By age fifteen, Trakr had lost the use of his rear legs, and was at the point where Symington had to hoist the dog's hind end with a towel and waddle behind him on their trips outside, or rig him up to a set of wheels, allowing the dog to propel himself using his front legs alone. Trakr's days were running out, and considering the amount of time, training, and affection Symington had invested in him, as both work partner and friend, it's no wonder that thinking about the possibility of life without Trakr would leave the sandy-haired actor despondent.

Symington's spirits rose significantly when he heard he was a finalist in the essay contest, partly because it would bring Trakr the attention he long deserved, more so because the prize for winning was beyond anyone's wildest dreams—the kind of miracle that, though home gardeners accomplish something similar with a sprig of ivy and a glass of water, most people once thought only Jesus could pull off.

"Ohmigod!" Symington's girlfriend could be heard screaming when the man at the door broke the news.

"Hello, are you Mr. Symington?" Lou Hawthorne asked.

"I certainly am."

"You're not just a finalist. You actually won the contest."

As winner of the "Golden Clone Giveaway," Symington would receive a free cloning of his dog, paid for in full by Hawthorne's biotech company, BioArts International. The month before, BioArts had held the world's first online dog-cloning auction, taking bids to clone five dogs. To head off criticism that the newly marketed service was something only the very rich could afford, Hawthorne came up with the essay contest, inviting dog owners to write about why their dog, above all dogs, should be cloned. Of all the heartwarming tales of "cloneworthy" dogs that poured in, Symington's stood out.

In the days ahead, video of a grateful and teary Symington would be posted on BioArts' website, Trakr's 9/11 heroics would be recounted in news media reports about the contest, and a small chunk of Trakr—after a biopsy—would be flown to South Korea. There, cells would be isolated and inserted into the harvested and emptied egg cells of anonymous dogs in heat. The reconstituted cells would be zapped with electricity, to spur them into dividing. Once they did, they'd be placed into the uterus of surrogate mother dog. If all worked out, a pup or pups would be born two months later—genetically identical to the original.

If Trakr could hold on long enough, he would become the first dog in America to meet his clone.

■

Somewhere in America—perhaps the Chateau Poochie Dog Hotel and Spa in Pompano Beach—there's a dog getting a massage right now, maybe complete with aromatherapy, followed by a "pawdicure" or some quiet moments of canine meditation in the Zen Wellness Center.

Somewhere, there's a dog scavenging alleyways, nosing off trash-can lids in search of his next meal and avoiding the local "Animal Control" agency whose job is to incarcerate and, if need be, euthanize unclaimed, unwanted, surplus souls like him.

Somewhere—Beverly Hills, for sure—there's a small dog peering out the top of a designer shoulder bag as he's toted down a sidewalk, a fashion accessory with a heartbeat, a pampered papoose that serves to satisfy the motherly instincts of someone with no time for motherhood. He's probably more than content to play along.

Somewhere—be it an abandoned urban factory or a rural back lot—there are two dogs in a pit, biting into each other's flesh, fighting to the death. It's not because it's their instinct; it's because it's what their owners—be they wannabe urban gangstas, bored hillbillies, or NFL quarterbacks—have, through training, teasing, and torture, tasked them to do.

And somewhere, there's a purebred standing at attention in a show ring, not a hair out of place, waiting to feel the gentle cupping of a dog-show judge's hand around his scrotum, an obligatory "conformation" check to ensure that, whatever unseen genetic problems decades of inbreeding may have caused, the dog has an appearance, right down to the symmetry of his gonads, that sufficiently complies with kennel club standards.

Whatever we try to make them, whatever we ask, dogs go with the flow. It's what they do best—despite the incredibly mixed messages we, over the centuries, have been sending them.

They've learned to adapt to domestication, industrialization, and urbanization. They came in from the wild, then let us work the wild out of them—at least that which interfered with their performing the jobs we selectively bred them to do. They happily complied when we shifted their primary role from worker to companion, consenting to collars and leashes, funny haircuts and sweaters, bandannas and bling. They gladly accepted our invitation to come live indoors, adjusting their toileting habits accordingly.

They have willingly—or, at least, it seems that way—filled the roles we have mandated for them, whether it be hunting assistant, herder, sled-puller, drug-detector, bomb-finder, searcher/rescuer, fighter, racer, gentle guider of the blind, fierce protector of property, substitute child, or simply the huggable antidote to loneliness. They've adapted not just to our basic needs, not just to the ridiculous extremes to which we sometimes go, but to our uniquely human peculiarities—our smothering love, our cruel abandonment, our neediness, our use, misuse, abuse, experimentation, and greedy exploitation of them, from our dictating how they live to our selfish refusal to let them die.

We, in large part, also decide whose genes will live on, who among them will breed with whom, particularly when it comes to purebreds. We do our best to ensure no hanky-panky among the rest, rendering their sex organs nonfunctional, then scolding them when they enjoy a brief moment of phantom humping, groin licking, or any of a host of other lingering instinctual behaviors that offend us. Bad manners, by human terms.

We've shaped them through centuries of selective breeding, turning the basic wolf into 167 breeds, by the American Kennel Club's latest proclamations, then, through inbreeding, exaggerating their trademark features to the point of caricature. We've designed teacup-size Chihuahuas, mastiffs almost too big to fit through the front door, shar-peis whose folds of skin are so pendulous they can't see, bulldogs whose heads are so large most can no longer be born without C-sections. We've taken the already cartoonishly elongated dachshund and stretched it some more—not thinking so much about the kind of back problems to which that might lead.

For reasons honorable and not so honorable, with results good and bad, we've engineered their looks, behavior, hair length, hair texture, the length of their muzzles, the size of their ears, the tone of their bark, the arc of their back, the curve of their tail—everything down to the amount of space that is "right" between their eyes.

For centuries, we've exercised our presumably God-given "dominion" over dogs, implementing a regimen of control that starts well before the cradle.

As of 2008, it also extends beyond the grave.

It took eight years of trial and error, at two universities, on two continents, to clone the world's first dog; only two more years to put that service on the market—if even that. In a way, the marketing of dog cloning began before a dog was even cloned, with the creation of gene banks in which hopeful dog owners could store their pet's DNA, with the promise that the technology was just around the corner.

It took a while to get to that corner, but at the end of 2008, the first commercially produced canine clones were picked up—though not without some hitches—by their owner, Bernann McKinney, who, in taking receipt of five genetic replicas of her pit bull Booger, more than two years dead, became the first pet owner to have a dog cloned for the singular purpose of recapturing love lost.

By the middle of 2009, two companies were cloning dogs. They had delivered more than a dozen clones, and hundreds more orders and inquiries were coming in from pet owners. The sort of thing that, as late as 2000, was still ensconced in the realm of science fiction—

remember RePet, the one-stop mall cloning store in the movie *The Sixth Day*?—had arrived in the real world. The sort of thing one couldn't put a price tag on suddenly had one: it was $150,000. The sort of thing that, had it happened with humans, would have been deemed horrifying and likely snuffed out at once, happened in the world of dogs; and, with the exception of the hand-wringing of a few animal welfare groups, it was pretty much taken in stride.

What also seemed lost on humans was the tremendous incongruity of it all: taking an animal whose single most admirable, endearing, and exemplary trait is its ability to live joyously in the moment, and thrusting its cells, unasked, into an uncertain future.

Given all the control we've exercised over the species, given centuries of selective breeding, commercially cloning them was merely a logical next step, say those providing the service. But it was, in many ways, a giant one. In the quest to clone dogs, the technology's quick transition to the marketplace, and the public's acceptance of it, man quietly crossed a threshold—or maybe two or three—taking a controversial technique, only ten years old and still restricted to experimental lab animals and livestock, applying it to the world's most beloved family pet, and making it available to the public, albeit for a rather handsome fee.

The cloning of dog brought us a step closer to what much of the world fears—human cloning—and it did so not just by advancing the science but by advancing the social acceptance of clones. Just as some have predicted that seeing cloned babies might melt public opposition to human cloning, seeing cloned puppies produced reactions closer to "Awwww" than "Eeeek." It's hard to look into the face of a puppy and see either evil monster or soulless doppelgänger.

But what we didn't see in the choreographed press conferences or the company-made videotapes of master-clone "reunions" was the sad, surreal, and sloppy story behind dog cloning's ten-year journey from billionaire's idle dream to purchasable commodity—one that allows those who can afford it to tell death, "No thanks," at least when it comes to our dogs.

We didn't see the massive harvesting of eggs on two continents, or the deformities, stillbirths, and deaths—all of which were routine in the quest to clone a dog but not required to be reported by scientists whose research was privately funded. We didn't see the frustration of squinty-eyed scientists trying to solve the microscopic mysteries of dog eggs, or the behind-the-scenes infighting over whether and how to market the service, or the public relations maneuvering by the cloning companies, all of which proved adroit at manipulating more than genes. We didn't see the con men, scammers, and cult leaders who seized on the wondrous scientific advancement to further their causes and feed their bank accounts. In fact, nearly all the futuristic concerns that have been raised about human cloning—those worst-case scenarios put forth by bioethicists and science fiction writers alike—surfaced in the joint corporate-collegiate quest to clone dogs and market them.

Like most good dog stories, the saga of dog cloning probably reveals more about humans than canines—our innate need for power, fame, money, and love; our thirst for instant, or at least speedy, gratification; our uniquely human refusal to accept the finality of death, and our overwhelming desire to control the world in which we live. And, like most good dog stories, it likely contains some lessons for those who navigate on two legs, the kind of wordless tutelage dogs have been offering for centuries—if only we'd listen.

With the cloning of dog, man may have crossed the final barrier to cloning humans. Clearly, he reached the pinnacle of his dominion over the species—the ability to fetch at least a semblance of his "best friend" back, not just from the jaws of death but from deep in its gullet. It's as easy—once all the microscopic intricacies, egg-timing issues, and cellular subtleties were figured out—as replacing the contents of a cell, zapping it with electricity, and giving it a womb in which to grow.

Two months later, with nature having picked up what man started, there's a living clone—not, usually, a monster; not, always, a mirror image; and not, certainly, a reincarnation of the original pet, as many clients admit is their hope. Instead, it's a new being, grown from the cells of an old one, taking shape in the womb of a surrogate mother until—eyes closed, paws drawn up to his muzzle—he's ready, most often with some surgical coaxing, to come into the world.

Here, boy.

Good dog.

2.

The Missy Mission

■ ■ ■

> To his dog, every man is Napoleon; hence the constant popularity of dogs.
>
> —*Aldous Huxley*

SAN FRANCISCO, CALIFORNIA

February 1997

The breakfast table had all the ingredients needed for the birth of an idea—three unrushed humans, one lounging dog, plenty of coffee, and the Sunday *New York Times*.

John Sperling, at seventy-six, was, as always, looking forward. He had accumulated more wealth than most people can imagine—$320 million in 1994 alone, when the University of Phoenix, his for-profit school for adults, went public. While watching his wealth multiply, he had started funding a handful of futuristic research projects, from growing crops in sea water to looking at ways to extend life expectancy. A man with little respect for limits that others, nature included, might impose, Sperling was especially drawn to those pursuits that, however unpopular, ventured into the unknown.

Joan Hawthorne, his longtime friend and lover, in her seventies as well, was looking back, formulating the next in her series of memoirs that, under her pen name, would recount her painful marriage, bitter divorce, brutal child custody fights, a battle with breast cancer—a disease she suspects she contracted from working in and around government weapons facilities—and quiet moments with her dog Missy.

Her son, Lou, was mostly just looking. At thirty-seven, he had a childlike curiosity, a vivid imagination, and an ongoing love for *Star Trek*. He'd graduated as a creative writing major from Princeton University, worked in software development and documentary film production, and was just back from a group motorcycle trip across India. The ride was to be part of a documentary called *Hell's Buddhas*. He had to scrap his movie due to lack of funding, but the trip went on. Hawthorne returned home to find himself between adventures, jobless, and with no clear course ahead.

Off to the side—close enough to scarf up any fumbled bacon, stray English muffin crumbs, or errant scrambled eggs—was Missy. At ten, the much-loved border collie–husky mix, adopted by Joan Hawthorne from a Northern California shelter as a pup, had slowed down a step or two but still maintained the remarkable energy and zest for life that her owner both marveled at and envied.

As for *The New York Times*, it was undergoing its usual Sunday-morning fate—its sections divided and passed around, their well-thumbed contents serving as fodder for the leisurely breakfast table conversation that came and went amid the random clink of cups hitting saucers. Among the newspaper's front-page headlines was this one: "Scientist Reports First Cloning Ever of Adult Mammal."

Everyone at the table had read the article, or at least its beginning:

"In a feat that may be the one bit of genetic engineering that has been anticipated and dreaded more than any other, researchers in Britain are reporting that they have cloned an adult mammal for the first time."

Dolly the sheep had been born at the Roslin Institute in Scotland, and the announcement of her birth was leading to the dawning public recognition that any animal could likely be cloned, even man. "It's unbelievable. . . . It basically means that there are no limits. It means all of science fiction is true," the article would quote Princeton biologist Lee Silver as saying. "They said it could never be done and now here it is, done before the year 2000."

The remark Sperling would make after perusing the article was lighthearted and offhanded—intended as a conversation starter more than a call to action. "Hey," he said, "we should clone Missy."

But an idle thought, when it occurs in the brain of a billionaire—especially a maverick, caffeinated billionaire, surrounded by supportive friends—often becomes more than an idle thought.

Lou liked the idea. Joan found it at least entertaining. As the conversation progressed, with each new thought serving to nourish the seed Sperling planted, the fanciful idea made the transition from pie in the sky to viable pursuit—one that, given Sperling's money, was possible, and, given America's long, deep, and steadily growing love affair with dogs, could even prove profitable.

By 1997, dog had become not just man's best friend, but, in many ways, his four-legged soul mate.

As the twenty-first century approached, dogs in the Western world were enjoying the highest status in their history. While they had been cherished, and even worshipped, over the ages, going as far back as ancient Egypt, something profound happened with dogs in the second half of the twentieth century—perhaps the biggest transition in the 15,000 years since they were domesticated.

They'd wandered into our villages way back then, probably lured by the scent of our garbage—step one of their evolution from wild wolf to man's best friend. At first, they were valued for the work they could perform, and bred for the jobs we assigned them. But they'd also take hold of our hearts, so much so that, by the mid-1900s, we became even more hospitable, inviting them indoors.

That overture led to more transitions—from worker to companion, from backyard denizen to roommate, from domesticated animal to emotional codependent. It happened virtually unnoticed—evolution takes place too gradually to be the "top story tonight"—but in just the past fifty to sixty years, dogs have moved up several rungs on the social ladder. They've entered our homes, hopped into our beds, and joined us, as near equals, on the sofa for some TV.

The relationship between the species is no longer based on food in exchange for work, or even food in exchange for love. Today, the relationship between pet and owner, though still symbiotic, has reached the point where the animal is not just being treated more like a human, but is most often considered a full-fledged family member.

Dogs are now most commonly given human names, as opposed to distinctly doggie ones. There are fewer Spikes, Spots, and Rovers; more Sams, Jakes, and Dylans. Half of Americans consider their dog an equal member of the family, according to recent surveys. A third of Americans allow them in their beds. More than 90 percent say they would risk their own lives to save their pet. In some large cities, dogs now outnumber children. Dogs are pushed in strollers, dressed in clothing, and dropped off at day care. They visit therapists, are given antianxiety medicine, and have birthday parties thrown for them. Though few owners provide for their pet to the extent of Leona Helmsley, the hotel heiress who bequeathed millions to her Maltese, named Trouble, more people are establishing trust funds to ensure that their pet is provided for after their demise.

In a world grown seemingly meaner and less secure, where humans connect to one another most often through the Internet and from a distance, dogs—through their touchable, physical presence and steady love—offer solace. Amid environmental fears, terrorism concerns, economic uncertainty, and frightening technology, dogs help us stay on an even keel, keeping our blood pressure low and, in some cases, maybe even our sanity intact.

Given all that, dogs have grown more popular every year. By 2009, in the United States, even amid a recession, we'd be spending more money on them than ever before, and drawing closer to the day when half of all homes had at least one dog. By then, their population—not counting the millions in shelters—was at a record high 77.5 million.

Missy, in life, was as loved as any of them.

Adopted by Joan Hawthorne as a four-month-old pup from a shelter near Santa Cruz, Missy was a mix of several undetermined breeds. Some thought she looked part wolf, others thought husky. Her behavior suggested there was some border collie in her. Some suspected there might even be some coyote in the blend. Joan Hawthorne compared her dog to a Rorschach test: "People saw the breed or blend they wanted to see."

Missy was frisky and impish as a pup, yet sweet and gentle. When she played, she played all-out, to the point of exhaustion. When she slept, it was sound and deep. And when she turned on the charm, humans would melt, just as Joan Hawthorne did when she first met her in 1987. Hawthorne was seeking a companion for her other dog, a fifteen-year-old shepherd-coyote mix named Liebe, who was losing the use of her back legs. "We decided to try to find another dog to cheer her and to make her life more interesting, whether or not she approved," she would later write in a remembrance of Missy.

"I began talking to her and she sat there sloppily, one leg loose, cocking her head, left, right, and I noticed that one erect ear had a dip, which I found somehow charming. I offered a howl to her and she raised her nose and howled at the roof. I barked at her and she barked right back, a low rich-toned, businesslike bark. I whined, she whined. . . . I put my arms around her and buried my face in her brown-black-white fur. She smelled so good."

Not wanting to make a rash decision, Hawthorne spent the night thinking about the dog. She arrived at the shelter the next day, thirty minutes before it opened. "She's waiting for you," an attendant told her. On the ride home from the shelter, she recalls, Missy—the name the dog came with—obligingly got in the backseat at her command, but then wriggled and contorted herself into a position where she could lay her head on her new master's shoulder as she drove.

When Liebe died, Missy was there to help fill the void. When Joan Hawthorne learned she had breast cancer, Missy was there to silently calm and comfort her. Missy was beautiful, intelligent, lively, and amazingly agile—the kind of dog, Lou Hawthorne would later say, "that only comes along once in a lifetime."

If the Sperling/Hawthorne household had its way, she'd be the first dog to come along twice or more—the first dog to be re-created, like the famous sheep, from one cell, fused, zapped, and implanted in a surrogate.

Before the breakfast dishes were cleared at Sperling's home, before

The New York Times landed in the recycling pile, a rough course had been charted: Sperling would foot the bill. Lou Hawthorne would do the legwork, investigating what it would take to put together a team of scientists to accomplish the cloning of Missy.

It was hardly the first time Sperling, a self-described rebel and risk-taker, had gone out on a limb. It wouldn't be the last time Joan Hawthorne, the daughter of former San Francisco newspaper columnist Marsh Maslin, tried to recapture the past. For Lou Hawthorne, it was a chance to do what he liked best—lead an effort that, much like the starship *Enterprise,* would "boldly go where no man has gone before," and, in the process of crossing that frontier, possibly lead to a multimillion-dollar business venture.

There were details to be worked out, obstacles to be surmounted, and a decade of trials and errors ahead. But, in concept at least, the cloning of man's best friend—a quest based partly on love, more so on money, and maybe mostly on the challenge of it all—had begun.

3.

Booger

■ ■ ■

It's not the size of the dog in the fight, it's the size of the fight in the dog.

—*Mark Twain*

MINNEAPOLIS, NORTH CAROLINA
1997

In a sleepy little town in the mountains of western North Carolina, midway between Cranberry Gap and Carpenter Bottom, there lived a woman with many friends. All of these friends had four legs.

By her mid-forties, Bernann McKinney had mostly given up on humans. With the exception of an often bumpy relationship with her parents—a school principal and his wife, who lived at the bottom of the hill, about a quarter-mile from their only daughter—she dealt with people only when it was necessary: on trips to nearby Newland for supplies, on visits to the vet, or when seeking hired hands to help with her animals.

Otherwise, she preferred the company of her horses and her dogs. Dogs didn't judge you, or turn on you, or abandon you. They just loved you unconditionally, and that was the kind of love McKinney demanded. She had an especially soft spot for pit bulls, those short, sturdy, square-jawed sorts—not actually any one particular breed—commonly used in dog-fighting operations.

Once seen as endearing enough to serve as RCA Victor's mascot, and as Petey, who regularly appeared in the *Our Gang* movies and the *Little Rascals* TV show, pit bulls had by the early 1990s developed some serious image problems as a result of their use by dog fighters and their popularity among urban gangs. Because so many had been trained to be so violent, pit bulls rose quickly to the top of the most-feared-dog charts, replacing Rottweilers, Doberman pinschers, German shepherds, and other breeds that, on the basis of the behavior of a few of their members—generally instigated by humans—had taken turns topping the list.

Pit bulls, partly because of the serious damage they could inflict, came to be seen as innately evil and not to be trusted, and the myth even developed that once a pit bull bites down, its jaws lock into place. The media fueled the flames, playing up the violent side of the breed. "Dog bites man" suddenly became news again—at least when the dog was a pit bull, or mistakenly adjudged such by police officers. As a result of becoming the most feared of dogs, they also became the most often euthanized when they ended up in shelters.

Maligned and misunderstood, abused and abandoned—that was the pit bull's plight, and it was one to which McKinney could relate. A reclusive sort by then, she was the subject of a few small-town rumors. She knew firsthand how easily a reputation could be trashed, and how hurtful that could be. Just as some people cross the street to avoid walking by a pit bull, there were those in her own community who adopted the practice when it came to her. Staff members at the local weekly newspaper remember drawing straws whenever McKinney would call, generally to alert them to an animal in need of help. Whoever drew the shortest one had to talk to her.

She sensed in pit bulls a kindred spirit—a dog that, like her, was fiercely determined, unrelenting, and living with a bad rap.

McKinney had rescued many dogs, a few of them pits, over the years, but none would have as big an impact on her life as did a mostly black, unneutered pit bull that she came across in the summer of 1996.

"I was driving along the road in the mountains and I saw this little stray pit bull going through garbage cans on the side of the road, looking for food. He was halfway slurping down a McDonald's wrapper . . . this beautiful little dog with butterflies flying around his nose in the middle of all this garbage."

McKinney opened her car door, coaxed him in, and, suspecting he'd likely be euthanized if she took him to the local shelter, brought him home and named him Booger.

There, on the horse farm McKinney managed for her father, Booger would join her menagerie, which also included a guard dog her father had given her to keep out trespassers. Tough Guy, as he was called, was a mastiff, ordered from Philadelphia—an enormous dog trained to attack on command. He'd nipped her father a few times, but was as sweet as he could be to her. He'd been taught to chomp down on an intruder's wrist and not let go until the "safe word" was spoken, at which point he would, in theory, release whoever was in his clutches.

The safe word was "kiss."

About a month after she brought Booger home, as the air turned hotter and heavy rains swelled the Toe River, which runs alongside her property, the bees appeared—the way they did every summer. McKinney was highly allergic to them. On a muggy day in August, when bees got into her house, she grabbed a magazine, rolled it up, and began swatting, trying to chase them outside.

Tough Guy reacted, lunging at McKinney and biting down hard on her wrist. When she freed herself and ran outside, he followed, knocking her down, latching onto the other arm powerfully enough to hit bone. As she rolled on the ground—the "safe word" never once entering her mind—he bit her arms, legs, and stomach, the blood that gushed out only seeming to make him more frenzied.

Unable to fend him off with her shredded arms, she screamed: "Help me, God! Help me, Jesus! Help me, Booger!"

Booger, she says, responded. Although he was inside, he somehow got out, ran into the yard, and pounced on Tough Guy. Tough Guy was nearly three times his size, but Booger's fierce attack gave McKinney time to get to her feet and stumble to her car.

Because heavy rains had made the bridge from her property to the road impassable, McKinney had to drive across the bumpy countryside, steering with her elbows, down the hill to her parents' house, where she leaned on the horn until her father came out.

He rushed her to a local clinic, where doctors attempted to stop the flow of blood and tried to make arrangements for a helicopter to fly her out of the mountains and into Winston-Salem, where a team of trauma surgeons would be ready to do the surgery deemed necessary to save her arms. When heavy fog prevented a helicopter from coming in, McKinney was placed in an ambulance for the two-hour drive.

A nurse sat by her side, speaking to her in Spanish until they arrived at Wake Forest University Baptist Medical Center.

There, a team of six doctors put her under anesthesia and began trying to repair her shredded right hand, three of its fingers barely attached; a left arm damaged up to the elbow; gashes in her stomach; and a right leg torn from kneecap to ankle. When she regained consciousness after the initial twelve-hour surgery—there'd be more later—doctors told her they had managed to save her arms, and possibly her mangled hands, but they told her she would probably have only limited use of them.

The first thing she asked about was the dogs.

Tough Guy, she learned, was dead. Booger was in pretty bad shape, found bitten, bruised, and with a four-inch-long gash in his hip. Her father told her that Booger had been taken to the vet, and would have to stay there for a couple of weeks, but was expected to survive.

To this day, McKinney isn't certain what triggered Tough Guy's attack. It could have been the bee sting he received earlier, or the steroids she gave him to treat it. She suspects he may have been the victim of an overdose, since she learned later that instead of the 10 milligram doses of prednisone she had requested, the pharmacy mistakenly supplied 100 milligram pills. It could have been that McKinney came too close, or even accidentally struck the dog while swatting at bees.

A necropsy on Tough Guy would turn out to be inconclusive. A blood vessel had burst in his brain, but the cause of that could not be ascertained.

After multiple surgeries, doctors managed to reconstruct McKinney's arms into "something that looked semi-human," she says. With physical therapy, they told her, she could regain at least some of the use of both hands. Sent home, McKinney spent another month bedridden, recuperating—as Booger, at her side, did the same.

"It was horrible, mind-boggling pain, and you sort of have a pity party," she says. "You feel sorry for yourself and in that initial stage you get down and out and depressed." Her Southern Baptist parents, while caring for her, offered little encouragement. "I remember being in bed and hearing my mom, standing out in the hallway, asking my dad, 'Is she always going to look that bad, is she always going to be that scarred?' . . . I was an only child, homecoming princess, a cheerleader, a beauty queen, and suddenly I was scarred from head to foot and they just couldn't handle it." Other than the visiting nurses who changed her bandages, and an aunt who would help her bathe, McKinney says, there wasn't a lot of help forthcoming. "I didn't have any human friends that cared."

But she did have Booger.

Although he had never received any training as a service or assistance dog, Booger took on the role, helping McKinney with all the once simple tasks that had become impossible for her, like turning doorknobs, taking off her shoes and socks, getting in and out of the bathtub.

"As I healed, he healed. He sat beside me. He pulled my wheelchair. He took off my shoes and socks, and got me soda pop out of the fridge. He would actually help me with my laundry. He would go into the dryer and reach in and get out a towel with his mouth. He knew the difference between a towel, a T-shirt, and a pair of jeans. He just seemed to understand. I never trained Booger to be a service dog. He just was one. And he never made a mistake."

To help her get in and out of the bathtub, she had a harness made for Booger with a handle she could grab hold of to get up and down. "If I fell down, then he'd come over and switch his rear end around, for me to grab onto his service harness, and he'd pick me up off the floor. He was my hands. He was my friend." For those tasks he couldn't help with, she figured out a way. "I didn't want somebody else brushing my teeth," she says, "so I had a hole drilled in the wall and stuck my toothbrush in it." For months, she brushed her teeth by moving her head up and down, back and forth, around the stationary toothbrush.

Sitting in bed and watching TV—something she did a lot of at first—she used her nose and mouth to push the buttons on the remote control, or to make a phone call. Generally, those were to Elliott Brown, a widower in California she had never met in person. They had met in an online chat room, and he was patient enough to wait for her bandaged hands to type the slow and garbled words that appeared on the screen. More often, they talked by phone. Brown, retired from Northrop Grumman, where he helped build B-2 bombers, had just lost his wife to cancer. He became McKinney's long-distance friend and confidant. He'd send her encouraging cards, usually with pictures of birds flying in the sunset. On the first Christmas after the attack, McKinney mailed him a Christmas card, which she signed by holding a pen between her toes.

Other than Brown, she saw Booger as her only ally. With Booger's assistance, McKinney says, she was able to return to managing the farm. While the use of her right hand was returning, Booger still routinely helped her get in and out of the tub, pulled off her shoes and socks at night, and assisted with doorknobs, which she still can't turn with her left hand. She continued her therapy, leaving Booger in the car, parked in the shade, during the sessions, and working at home with a robotic-type device that, once attached, mechanically moved her fingers for her, giving them the exercise they needed.

But more than any device, or any human, it was Booger she depended on most, Booger who provided her with emotional support, and Booger who gave her the incentive to keep going. "Booger taught me not to give up, not to feel sorry for myself," she says. "Booger made me feel like a whole person. . . . Booger taught me I could do anything I could do before—that I just had to figure out a different way to do it."

As her condition continued to improve, she began sharing with others what she saw as her dog's special ability to help soothe and heal, almost as if he were absorbing her pain and anguish. She and her dog would visit nursing homes, where Booger would sidle up against veterans who'd lost limbs in wars—sometimes, she says, bringing tears to their eyes. Once, in a shopping mall, Booger noticed a young girl in a wheelchair; he walked over and stood alongside it, as he had done when McKinney was using one. "He was like, 'Go ahead, grab the harness, I'll lift you out,'" McKinney says. "Her mother looked at me and cried. She said, 'Where do I get one of those dogs?' I said, 'Well, this is one God rented out to me from heaven for a short period of time.'"

Though her beauty had been marred by the attack, though some functions were lost, and though she had gained weight during the long period of being bedridden, McKinney found her self-esteem again through Booger.

"When I was out with Booger, everybody was just amazed by him. He gave me self-confidence. I think a service dog does that for anybody that's handicapped. Service dogs bring out the spirit and strength and outgoingness of the person they are serving. Suddenly you're as good as the next guy."

While going through more physical therapy and surgeries, including one in which bone was removed from her hip to replace the one lost from her wrist, McKinney continued to mourn the dog who died, as well—Tough Guy, against whom, even though he maimed her, she seemed to hold no grudge.

In 1997, McKinney went so far as to call Ian Wilmut, the Scottish scientist whose historic cloning of Dolly the sheep at the Roslin Institute had been accomplished the summer before but not announced until that February. When she asked him about the possibility of cloning Tough Guy, Wilmut politely answered her questions but told her dog cloning was still likely years away.

It was then, too, that McKinney first thought about someday cloning Booger, whose self-learned skills as a service dog were—to her, at least—dazzling. While there was nothing she could do for Tough Guy, she held out hope that technology would progress to the point where she would be able to clone Booger when his time came.

It shouldn't take long, she figured, for scientists to get from sheep to dog.

4.

Trakr

■ ■ ■

All animals are equal, but some animals are more equal than others.

—*George Orwell,* Animal Farm

HALIFAX, NOVA SCOTIA
1997

There's an old essay question—one that countless high-schoolers have pondered: Are heroes born or made? Are some of us destined for greatness because of a special brave-and-fearless hero gene in our DNA? Or are heroes made—made, in this case, not meaning manufactured but shaped by circumstances? Is there a hero in all of us—dog or human—that, given the opportunity, will rise to the occasion? For a German shepherd named Trakr, the answer was: a little of both.

Trakr was born in 1993, the same year his country of birth, Czechoslovakia, split in two, becoming the Czech Republic and

Slovakia. He was purchased from a Czech breeder of police and guard dogs and was fourteen months old when, in 1995, he became the first canine to go on duty in the Halifax, Nova Scotia, police department's newly formed K-9 Unit.

Trakr had the prime bloodlines, he had the training, and—as a police dog—he had more than a few opportunities to show what he was made of. On his first day on the job, he assisted in an arrest, to the surprise of everyone except maybe his handler and partner, James Symington.

Symington had grown up in Lower Sackville, Nova Scotia, outside Halifax, one of four children of a dockyard mechanic and his wife. Graced with rugged good looks, he put his teenage dreams of acting aside and went straight from college to the Halifax Regional Police Department, joining the force in 1988. He worked in the mounted division, harbor patrol, street crime, the SWAT team, and underwater search-and-recovery.

In 1995, he and another officer started the department's K-9 unit with Trakr and another German shepherd custom-ordered from the same Czech breeder. The Czech company, to ensure a good fit, went so far as to conduct tests, eHarmony style, to make certain that dog and handler would be compatible. If Symington and Trakr were any indication, the company knew how to make a match.

For Symington, it was love at first sight. As he put the dog through training, Symington found he was getting back at least as much as he invested—both as a reliable partner and as a trustworthy friend. Trakr, like many police dogs, resided with his handler. Between that and depending upon each other on the job, the two formed a deep bond. While Symington's family had dogs when he was growing up, they were family dogs, and he never saw any of them as being distinctly his. In that way, Trakr was his first.

After setting the pace with an arrest assist his first day on the job, Trakr would go on to find missing people, track down criminals, and help recover more than a million dollars' worth of stolen goods in his six-year police career. "He was just the perfect cop," Symington says.

He was an excellent goodwill ambassador, as well.

In their spare time, Trakr and Symington visited children in hospitals and made appearances before youth and church groups. Trakr was the first police K-9 to have his head shaved for "Cops for Cancer," a national initiative raising money for children with cancer. Trakr had his own page in a coloring book and his own trading cards. Trakr could collar a criminal one day, cuddle with disabled children the next—a rare combination in a police dog.

By 1997, he was already becoming a local hero. Four years later, he would become—to a limited extent—a national one. And seven years after that, retired and having lost the use of his rear legs, the German shepherd would be chosen as the most "cloneworthy" dog in the world, based on an essay recounting his heroics, written by Symington. Trakr would be "made" again—this time in the manufacturing sense of the word—partly because he was a hero, partly because, to the company behind it, it was good public relations.

America loves its heroes. It loves its dogs. But it especially adores a hero dog—a staple of television, movies, and literature even before Lassie hit the airwaves. Tales of brave and loyal dogs going above and beyond the call of duty—giving their all for the human race, or at least one of its members—are as old as the species. It's not a uniquely American phenomenon. Nearly every culture has its hero-dog stories, even in those places where dog is commonly a menu item.

In Korea, there are two versions of the legend that best exemplify the loyalty of an ancient, docile, and shaggy breed of native dog known as the Sapsaree. Both versions start off with an aristocrat who is heading home with his dog from a ceremony at which he over-imbibed. Staggering down a path along a river, he passes out. In one version of the tale, it's the ember from his pipe that starts a fire in the surrounding brush. In the other, the fire just ignites out of nowhere. In either case, the drunken aristocrat is about to be engulfed by the spreading flames when his dog leaps into action.

He jumps into the river, soaking up the water with his locks like a dry mop, then jumps back on land, shaking his wet fur. The dog repeats the process again and again, until the fire goes out. How many soakings/shakings it took isn't specified, but apparently enough to take a toll on the dog. After saving his master—or so the legend goes—he dies immediately of exhaustion.

Out of appreciation, the aristocrat erected a stone monument to his dog, which still stands today in Seongsan County in the province of Gyeongsang. It's just one of many images paying homage to the Sapsaree, repeated in sculptures, called *haetae,* across the rural South Korean countryside, where many still eat dog, though not usually the Sapsaree.

In Japan, there's the more documentable legend of Hachiko, an Akita born in the prefecture of Akita in 1923. He moved with his owner, Eisaburo Uyeno, to Tokyo in 1924. Every day, Hachiko would accompany his master, a university professor, to the Shibuya train station and watch him leave for work. Every evening, the dog would be patiently waiting for him at the train station for the walk home. After about a year of the routine, Uyeno became ill at work and died. Hachiko, though his care was undertaken by others, still went to the train station every day and waited for his master—for nearly ten years. He died in 1935, at the train station, while waiting. A statue of him now stands there; his mounted remains are kept at the National Science Museum of Japan in Tokyo; and his story has been retold in numerous genres, from children's books to, in 2010, a remade-movie version, set in Rhode Island and starring Richard Gere.

The same year Hachiko began his train station vigil, a diphtheria epidemic appeared destined to devastate Nome, Alaska. The only serum that could halt the outbreak was a thousand miles away, in Anchorage. Air travel, a relatively modern technology at the time, wasn't possible, because the only airplane capable of quickly making the trip wouldn't start. So teams of sled dogs were assembled to transport the medicine. A Siberian husky named Balto led the team that ran the last stretch.

While other dogs led teams longer distances under harsher conditions, it was Balto who became the celebrity. A statue of him was erected in Central Park the same year. Balto attended the unveiling, then went on to further fame on the vaudeville circuit. But when Cleveland businessman George Kimble saw the conditions in which vaudeville dogs lived, he decided the hero dog deserved more, and began a campaign to bring Balto to a zoo in Cleveland.

Balto lived out his life in what was then called the Brookside Zoo. When he died in 1933, his pelt was mounted by a taxidermist and donated to the Cleveland Museum of Natural History. He has remained there, except for a five-month stay at the Anchorage Museum of History and Art, approved after the Alaska legislature passed a "Bring Back Balto" resolution. Today, more than eighty years after his feat, Balto, a singular dog, is remembered with a statue in Central Park; a "stuffed" Balto on display in Cleveland; and the grueling, 1,100-mile Iditarod sled-dog race in Alaska, an annual event that, while it commemorates a life-saving feat, often results in injuries, and sometimes deaths, among the dogs competing.

In modern times, heroic dogs seem more likely to come to attention than ever, whether they work for law enforcement, serve as assistance and therapy dogs, or are just plain old civilians. In recent years, the reported feats of family dogs include notifying owners of house fires in time for them to get out, dialing 911, saving children from cougar attacks, and even, as one woman in Maryland claimed, performing CPR.

While the news media once dictated which canine heroics were worthy of fame, it's now possible, thanks to the Internet, for Web-savvy dog owners—some of whom go so far as to hire public relations firms—to shape a legacy for their pets online, ensuring them a place in history, and sometimes, in the process, giving themselves a shot at book or movie deals.

Trakr, in later years, would see James Symington and his wife—his biggest boosters—avail themselves of such services, but he also had an impressive law enforcement résumé. It was one that handlers of thousands of other search-and-rescue, bomb-sniffing, and police dogs could have compiled and distributed via blogs and press releases, were they so inclined—but with one major difference after 2001. Trakr, Symington says, found the last survivor of 9/11.

Symington rarely hesitated to step into the limelight with his dog—throughout Trakr's career and after it ended. When Trakr retired, in May 2001, there was a public ceremony at a local high school, where Trakr received a medal and a plaque—all captured by the local media. When Trakr and Symington returned from their unauthorized trip to assist with the 9/11 rescue effort, it was a friend of Symington's who informed the local newspaper about their heroic feats, on the eve of the department's decision to take disciplinary action against Symington.

Symington, his yen to act having returned, was already on thin ice with the department for having taken sick leaves to work on movie sets. When movie production companies came to Nova Scotia to film, Symington was generally able to find work, as either a stuntman or an actor. In 2000, he worked as both in *The Shipping News*. The next year, he worked as a Harrison Ford's body double in the movie *K-19: The Widowmaker*.

After Trakr's retirement, Symington took a medical leave, claiming at first an elbow injury, and later job-related stress. Symington said that stress, in large part, was a result of the department's plans to euthanize Trakr and all other retiring police dogs. Halifax police officials deny that any such policy ever existed.

Amid continuing disputes and litigation with his department, Symington, taking Trakr with him, relocated from Canada to Los Angeles in 2003 for an acting career that would lead to soap opera roles in *General Hospital*, *Days of Our Lives*, and *The Young and the Restless*, and parts in a handful of movies, including *Cheerleader Massacre 2*.

Whenever possible, he'd share the spotlight with Trakr, whose health by then was beginning to deteriorate, prompting Symington to bank the dog's cells in 2005—before the cloning of dog had been achieved. Once it was, and became commercially available, Symington wasn't in a position to afford it. When a contest for a free dog-cloning was announced, he sat down and wrote his winning essay, recounting Trakr's heroics at 9/11. It won easily.

Fifteen years after Czechoslovakia divided, Trakr's harvested cells would do the same in a laboratory in South Korea—after being placed inside an egg from a donor dog, zapped with electricity, and inserted into another dog, who would serve as surrogate.

It would be, in the view of some, the first cloning of a hero. And if anybody deserves to be cloned, who more so than a hero?

In the endless debate over human cloning, the question often emerges of just who, once the technology is achieved, would get to avail themselves, or their cells, of the service. Whom would we clone, and why? Albert Einstein or Michael Jordan? Mozart or Michael Jackson? Mother Teresa or Lady Gaga? Who deserves to have his or her genes repeated? Would it be those who excelled in the sports arena, the laboratory, diplomacy, the arts? And what's to prevent somebody from sneaking a Hitler into the mix?

As it got off the ground as a commercial enterprise, dog cloning's earliest clients would include at least one athlete—a racing greyhound—and a few dogs cloned for law enforcement and medical uses, but by and large it wasn't about duplicating the most talented, the most endangered, or the best-looking.

Instead, the first pet owners to have their dogs cloned would be those with the fattest bank accounts, a tendency to want to be in control, and whose backstories—at least on the surface—went the furthest in adding a personal and heartwarming spin to the otherwise cold science of dog cloning.

5. Sperling's Dogma

■ ■ ■

> When a man's best friend is his dog, that dog has a problem.
>
> —*Edward Abbey*

COLLEGE STATION, TEXAS
1998

The man behind dog cloning in America—or at least the man whose money was behind dog cloning in America—was himself the result of an unwanted pregnancy, born in a log cabin in the Missouri Ozarks to, by his own account, a "suffocating" Scottish-Irish mother and a shiftless "ne'er do well" father. His past—though he and others say it lingers in his difficulty forming close relationships with people—is, in most other ways, well behind him, as evidenced by, among other things, his Scottsdale mansion and his private jet.

That jet landed in the middle of Texas in 1998. Its passengers—Sperling, Lou Hawthorne, and a dog named Missy—were picked up and whisked to Texas A&M University, where a team was standing by to gather cells from the dog.

If there was a better place to get Missy cloned than Texas's oldest public university—a school that takes its football, its canine mascot (Reveille), its military tradition, and its agricultural mission seriously—Hawthorne couldn't find it. And, after being dispatched to find the institution best equipped to clone Missy, he had made contact with at least twenty.

As Sperling's point man, Hawthorne had met A&M scientist Mark Westhusin at a Transgenic Animals in Agriculture conference in Tahoe City, California, the year before. "He was running around interviewing different people, trying to figure out who might be interested in taking on a project to clone a dog," Westhusin recalled. "He talked to several people. He ran into me and we sat down and talked, and I told him, yeah, I might be interested in exploring the idea.

"He told me that the person he represented—I didn't know who it was at that time—had a lot of money and was really interested in pursuing this. I told him if he really wanted to make a significant effort it would probably cost a million dollars a year and you'd probably have to spend four or five years just to get started."

A&M was one of at least three universities that Hawthorne invited to submit proposals for the project. Upon being selected, the research team of the veterinary school at Texas A&M received an initial $2.3 million from John Sperling, whose connection to the project was being kept secret at that point, at least from the general public.

Other than his initial trip to Texas A&M, Sperling would stay in the background, even once his identity became known, leaving the details to Hawthorne.

The collaboration was formalized, but it would prove to be a rocky one. "Lou and I were always back and forth," Westhusin says. "He had crazy ideas—some of them I'd go along with and some of them I'd just say, 'Nah.' It was just his way." The first one came before Missy even arrived on campus, and struck Westhusin as ill-advised.

Hawthorne suggested that the military school's renowned corps cadets be on hand for Sperling's arrival on campus, holding signs that said WELCOME MISSY TO AGGIE LAND. Westhusin favored a quieter approach, one that wouldn't attract either animal rights protesters or news media.

Arriving on campus, the two men and the dog received an official salute from a few members of the school's venerated Corps of Cadets—a toned-down version of the reception Hawthorne had envisioned, and without signs. "I just thought it was kind of silly," Westhusin says. "I think I decided we didn't have time to make signs."

Missy was whisked to a veterinary school lab, and a sample of her tissue was taken, from which the cells needed to clone her could be isolated and treated. Dubbed "Missyplicity," the project at Texas A&M would use roughly the same techniques Ian Wilmut had employed to create Dolly the sheep to build a brand-new Missy, or possibly more. It all went cleanly, quietly, and quickly, to Westhusin's relief, and, after collecting the tissue, they all went to Jason's Deli, where Missy sat next to the table as Hawthorne and Sperling fed her broccoli from the salad bar.

It wasn't until several weeks later, when an official announcement of the project came out, that controversies—minor ones—arose.

Westhusin gets his salary and research funds through two departments at the sprawling university—the veterinary school and the Texas Agriculture Experiment Station. It was the latter that approved the contracts for the Missyplicity Project and issued a short announcement about it, which came as a surprise to Ray Bowen, the provost of Texas A&M, who was unaware the veterinary college had undertaken the cloning of a dog.

Bowen initially said the contract was invalid and had lawyers review it, but eventually, with minor changes, it was approved.

"There was never any indication from Dr. Bowen or anybody that they didn't want us to take the thing on and try to do it, at least that I heard," Westhusin says. "No one came in and said, 'You can't do that,' or tried to stop me from doing it. But it did cause trouble amongst the administration, and it led to some changes in the way things happen around here. I just tried to keep my head low and hoped it all would work out, and it finally did."

Sperling had chosen, at least initially, also to keep a low profile in connection with the project—uncharacteristic behavior for a man who has never been averse to controversy and confrontation. The Missyplicity Project had all the ingredients Sperling savored. It had never been accomplished. It had edge. It was taking new science to new places. It, if successful, would give him an upper hand over nature, if not time itself, by taking life—a dog's life, at least—beyond its normal boundaries, and in the process perhaps cure Sperling of what has ailed him since childhood.

Sperling has been poor enough to reportedly steal a chicken, when he was a teenager, so he and his mother could eat. He has been rich enough to orchestrate the direction of scientific research. But, as both, he has been bored. Boredom followed him through his hardscrabble childhood, long stretches of which he spent bedridden due to illnesses and a heating stove explosion that left his face and hands covered with blisters. Boredom led him, when he was working as a stock clerk, to join the Merchant Marine. Boredom brought him to academia, then—when he got numbed by all its layers of bureaucracy and theory—sent him on the road to start his own university.

"I take action whether it is prudent or not, even when doing nothing seems to be the wisest course," he wrote in his autobiographical tome *Rebel with a Cause.*

In that book, Sperling portrays his upbringing as both miserable and meager. His father couldn't keep a job, and whipped him regularly. His mother's overbearing nature left him never again wanting "to have anyone with a hold on me. At whatever cost, I try to stand alone rather than be beholden to someone to whom I must acknowledge superior status."

At fifteen, when his father died in his sleep, Sperling says he celebrated with abandon. "I could hardly contain my joy. I raced outside, rolled in the grass squealing with delight. There I lay looking up into a clear blue sky, and I realized that this was the happiest day of my life. It still is."

As a teenager, Sperling moved to Oregon, graduated from high school, and worked briefly as a stock boy before signing up for a two-year stint with the Merchant Marine. On the seas, he fought his dyslexia and learned to appreciate books. He went on to work at a shipyard to pay his way through Oregon's Reed College. He enlisted in the Army Air Corps and, through the GI Bill, continued his education. He married his college sweetheart, then left her behind, saying good-bye in a note when he went off to graduate school at the University of California, Berkeley.

He went to Cambridge University and earned a Ph.D. in economic history, then taught at several colleges in the Midwest before ending up at San Jose State in 1960. He married again, and had a son, but under what he described as the "strains of cohabitation" Sperling procured living quarters of his own. It was about that time an old friend named Joan reentered his life. She was the wife of one of his graduate school friends. Sperling's wife and Joan's husband had been Fulbright scholars together in Munich, where Joan worked at Radio Liberation from Bolshevism, which Sperling describes as a CIA operation.

In Northern California, Joan had two children and, according to Sperling's autobiography, was coping with "the inanities of university society and her husband's serial infidelities." Sperling had a son and a less-than-ideal marriage as well. Sperling and Joan shared an interest in early childhood development, which they'd talk about for hours while taking the children on play dates. They also shared a disdain for the stuffiness of academia.

"Jack was boring, like all professors . . . infatuated with his theories and their articulation," Joan Hawthorne would later write in her memoirs, using the pen name Candida Lawrence for herself, the pseudonym "Jack" for Sperling. He was seen as "gruff, black, cold, Machiavellian, lawless, witty, earthy, profane, tender, tough, ruthless, manipulative, self-made, mocking, devious, remote, brave," she wrote.

"'Talk, I can always talk . . .'" she quotes him as saying one fateful night, over cheese and crackers. "'Whatever creativity I had before grad school was destroyed, and I am still incapable of independent thought or action. But there's one thing I'll tell you. The stuffiness of academic society I'll no longer tolerate. I don't like professors, their wives, their children, their casseroles. They spend all their money on microfilm and books and eat stuff that rots their guts, if they have any left to rot. Professors are impecunious, ungenerous, and pecksniffish.'"

That night, according to both her memoirs and his autobiography, they slept together for the first time.

As Joan went through a nasty divorce, lost legal custody of her children, and finally absconded with them to start a new life under an assumed name, Sperling would stay by her side—not physically, but emotionally and financially. Along the way—though they never cohabited permanently and his visits were sporadic—he would become the closest thing Joan's children had to a stepfather.

And a billionaire, to boot. The seed that would grow into the adult-education empire that is the University of Phoenix was planted in 1972, while Sperling was teaching at San Jose State University.

Sperling's reputation at San Jose State, in his ten years at the institution, had lost some luster by then. A union organizer, Sperling had pushed a year earlier for professors to go on strike. A quarter of them did, walking off their jobs at San Jose and San Francisco state universities for thirty-one days. They gained nothing, and nearly lost their jobs. Some of his colleagues harbored a grudge about that. Others viewed Sperling as controlling, contrary, and confrontational—a man who, rather than shying away from fights, at least those of words, was fueled by them.

He had already shown by then that he was not afraid to shake things up. There was, for example, the time his class held an Earth Day demonstration, laying to rest a yellow 1970 Ford Maverick, fresh off the assembly line. It was the original Earth Day, held the year after Woodstock and the moon landing. The idea to hold a funeral for a Ford had been conceived in Sperling's Humanities 160 class as a way to spotlight the role the automobile played in polluting the earth's atmosphere.

While Sperling, acquaintances say, has always been one who

prefers to hold the reins, in Humanities 160 he let the students take control. That was the whole idea behind the class: to allow students to immerse themselves in an issue, take action, and see the consequences. While the university administration played along, not all of San Jose State's students did. Students from the black studies program voiced outrage about white students buying and burying a brand-new car to show their support for improving a public transportation system most white students weren't using anyway. On the day of the burial, black students began taking their anger out on the yellow Maverick, using their hands and feet, and then baseball bats, to bash the vehicle. Eventually, as planned, the Maverick rolled down a ramp and landed with a crunching thud in its grave, adjacent to the student union, where it was covered with dirt by bulldozers. The class went on to campaign for a local ballot measure to upgrade the city's public transportation system. When it passed, and a new public transportation building was constructed with the funds it provided, the yellow Maverick was resurrected, compacted into a cube, and incorporated into the cornerstone of the structure.

All in all, having never burned any gas, the yellow Maverick got some great mileage. It brought attention to how our dependence on technology—in this case, the internal combustion engine—was polluting our planet. It showed students in Sperling's class that, as Crosby, Stills, Nash, and Young were singing that same year, you can change the world. And for Sperling, the class reconfirmed that the way to effect change was through bucking convention, disregarding all naysayers along the way.

"Without that lesson, I could never have become a successful entrepreneur," he wrote in *Against All Odds*, one of the four books he has written, two of them autobiographies. "The lesson was simple—ignore your detractors and those who say that what you are doing is wrong, against regulations, or illegal."

The University of Phoenix was built on that same foundation.

In 1972, Sperling was asked to oversee a federally funded series of classes to help police and schoolteachers better deal with juvenile delinquents. Impressed by their eagerness to learn, and noting the lack of alternatives for workaday adults who wanted to further their education, he presented San Jose State officials with a proposal for an expanded adult-education program.

When it was rejected, he took the idea to the University of San Francisco, a financially troubled, Jesuit-run university that, seeing his idea as a money-making opportunity, added the for-profit piece to Sperling's proposal—a concept that stunned academic purists and led to criticism that the program's stripped-down version of education amounted to little more than a diploma mill.

In 1976, facing threats to the school's accreditation, Sperling moved the operation to Arizona, where he would relaunch under a new name, the University of Phoenix. Critics called it assembly-line education, McUniversity, the anti-college. But over the next two decades, fighting for accreditation in each state along the way, the adult degree program Sperling started with $26,000 in savings and eight enrolled students would branch into thirty-seven states, Canada, and Puerto Rico, with campuses—sometimes little more than rented space in buildings—in more than two hundred locations, including, eventually, that biggest location of all: the Internet.

On the forefront of online education, peddling "real world" learning as opposed to hazy theories, and fighting state licensing agencies at almost every turn, Sperling persevered. When he took the university's holding company, the Apollo Group, public in 1994, he became a millionaire overnight, a billionaire within a few more years.

With a billion dollars, changing the world becomes something you can do a lot more than sing about. Sperling found himself entering his golden years not with more money than he knew what to do with—there's no such thing—but with an amount vast enough to seriously pursue some pet projects.

First, he sought to bring an end to the war on drugs—a wasteful practice, in his view, which amounted to little more than a welfare program for police officers and prison guards. Teaming up with fellow millionaires George Soros and Peter Lewis, Sperling—who admits to smoking marijuana during a bout with prostate cancer in the 1960s—began bankrolling state ballot initiatives, starting in Arizona, to decriminalize drugs and replace prison time with treatment.

From there, he moved into saltwater agriculture, sinking $10 million into a giant saltwater-irrigated farm in Eritrea, on the eastern coast of Africa, through a company called Seaphire that hoped to both fight world hunger and make a profit at the same time.

His funding of research to extend the human life span—motivated at least in part by his own desire to live longer, he admits—led to the establishment of Kronos, a company that operates an antiaging clinic based in upscale, retiree-heavy Scottsdale, Arizona. It provides a series of exhaustive medical tests for clients to identify what's most likely to end their life, and a nutritional, drug, and vitamin regimen to forestall it. It, too, has begun to turn a profit.

"Implacable opportunism, joy in conflict, and getting a thrill from taking risks" are the three behaviors that served him best in his days as a union organizer, the richest man in Arizona wrote in his autobiography. Forty years later, with his dog-cloning attempt, he got to exercise all three—though, at least initially, he chose to do so anonymously.

The contract prohibited Texas A&M from saying who had provided the money. The owners were a Bay Area couple, some reports said; a mystery millionaire, according to others. In September 1998, though, *The Dallas Morning News* reported that the man behind the money might be Sperling. While initial reports portrayed Missy as his dog, reporters tracked down Missy's dog license and found she was listed as residing at property owned by Sperling in Watsonville, California, and registered to an owner named Joan Hawthorne. When contacted by a reporter by phone and asked about Missy, the newspaper said, Joan Hawthorne denied any knowledge of the dog and hung up. Lou Hawthorne was mum as well: "I have nothing to say to confirm or deny the identity of the clients," he told the newspaper.

In reality, Sperling's connection to the dog, like his connection with Lou Hawthorne, was more as unofficial stepfather, occasional visitor, and financial supporter. There was a bond, but Sperling's relationships, by his own admission, are part-time propositions that take a backseat to whatever, at the time, he is trying to accomplish.

"He's entirely engaged and excited by what he does, but I wouldn't necessarily say he's the happiest guy," Sperling's lone child, Peter Sperling, told the Portland *Oregonian* in 1998. "He'd love to have closer personal relationships, but because of his childhood, I think he's afraid to get close."

Sperling has said he plans to leave his money to research upon his death, assuring that—even after he goes the way of the yellow Maverick—he'll have a continued impact on science. It has been estimated that his wealth, if used to create an endowment, would generate at least $150 million a year—for eternity.

Missy's shot at eternity, or at least having her genes live on in a duplicate dog, got under way as soon as Sperling, Hawthorne, and the dog left the university—the same day they arrived—detouring around a construction project under way at Kyle Field, the football stadium.

As it turned out, at the same time Texas A&M was embarking on creating the dog of the future, the school was digging up dogs of the past.

The school was adding more seats to the stadium's end zone, a project that required unearthing the four school mascots who had been buried there—all with their heads pointed so they might have "a view" of the scoreboard. Texas A&M was on its fifth Reveille—all since the third being collies—at the time of Sperling and Hawthorne's visit. The first Reveille was a black-and-white mutt, struck on the highway by a carload of cadets. They brought her to campus to recuperate in 1931, and the next year, Reveille—so named because she would howl when the morning bugle call was sounded—was officially designated as mascot.

In 1999, all four buried Reveilles—and the fifth, who died during construction—were placed in new graves in an adjacent plaza on the other side of the stadium. A new outdoor scoreboard was built, so that the buried dogs might still have a view, and a memorial plaque was placed in the plaza, which honors the eternally resting Reveilles.

It reads: REVEILLE I, AND THE REVEILLES THAT FOLLOW HER, WILL ALWAYS HAVE A SPECIAL PLACE IN AN AGGIE'S HEART AND SYMBOLIZE THE UNDYING SPIRIT OF TEXAS A&M.

6.

The Bullheaded Cowboy

■ ■ ■

> Never approach a bull from the front, a horse from the rear, or a fool from any direction.
>
> —*Old cowboy saying*

LA GRANGE, TEXAS

August 1999

As a teacher, wrangler, and clown—as one who faced both raging rodeo bulls and angry adolescents—Ralph Fisher knows the benefits of proceeding with caution, the importance of seizing control of a situation, and the folly of thinking you're really in charge. Some things in the universe are out of your hands. You live your life; you take your chances.

In the 1960s and 1970s, Fisher was a schoolteacher in Houston, working with students who—long before either ADHD or Ritalin came into existence—were more likely to be referred to as "dang troublemakers." One dealt with them as best one could. "I used my training with animals," Fisher says. "It just seemed to fit right in."

When not in the classroom, Fisher worked as a clown in what was the nation's most pitiless rodeo, the Texas Prison Rodeo in Huntsville, where, in one infamous event called "Hard Money," prison staff would tie a bag containing $1,000 to the horns of a bull, and watch as inmates tried to grab it without getting gored or trampled.

And he's the guy who, every four years, loads his wagons with longhorn steers, Brahman bulls, armadillos, and other assorted creatures from his Texas ranch and hauls them to Washington to put a little more *yee-hah* into the inaugural bashes sponsored by the Lone Star State.

As founder and proprietor of Ralph Fisher's Photo Animals, Inc., the soft-spoken, slightly built former rodeo clown has sent his animals to Super Bowls and All-Star Games. They've appeared in movies, on TV talk shows, and at countless corporate events, for which Fisher provides a variety of entertainment options—from armadillo races to getting your picture taken atop a bull.

"With the world's largest herd of saddle broke tame Texas longhorns and Brahmans, Ralph Fisher's Photo Animals are sure to be a crowd pleaser . . ." his promotional materials read. "With one million dollars in liability insurance, beautiful animals and state of the art photo equipment, we furnish your clients and guests with a lasting memory of your Texas event."

Lasting memories are another thing Ralph Fisher can tell you a thing or two about. When the star of his traveling show took ill—a huge but amazingly passive Brahman bull named Chance—Fisher sought to have him cloned.

Fisher's hope was that researchers at Texas A&M's veterinary school—the same scientists who had just started working on cloning a billionaire's dog—would be able to bring him the same animal again. Those hopes, and a few other things, would be punctured in the years that followed.

The cloning of Chance at Texas A&M University in 1999—in addition to being the institution's first cloning, the state's first cloning, the first steer cloned, and the oldest animal ever cloned—could also be considered the first cloning of a pet.

For Chance—though he was a source of income for Fisher; though he was of a species not generally cuddled, and notorious for their unpredictability; though he was just a few days from becoming hamburger meat when Fisher picked him up—was, more than anything else, a family member.

Chance was purchased at an auction in 1984 by a cattle-buying friend of Fisher's who thought that, despite the bull's tremendous size, he might make a good addition to Fisher's traveling menagerie of tamed bulls—Texas longhorns, mostly, who were placid enough to be petted and tolerate tourists climbing atop them for photo shoots.

"The old bull came up and leaned against the fence so he could be scratched, and the guy scratched him, and he thought of me and he bought him. I said, 'Oh no, there's no chance that big Brahman is going to be tame.' It's just very unheard of. I didn't go see him. But he called me a month later and said, 'I can't keep this thing forever.'"

Back at Fisher's ranch, Chance joined Tumbleweed, the Texas longhorn Fisher started his traveling show with, and whose hooves Fisher doubted Chance could ever fill. To his amazement, Chance was even tamer than Tumbleweed. "We started riding him right away," Fisher says. "We didn't put a ring in his nose until we had him two or three years. That's where you get your real control. You know, ultimate control."

On top of his tameness, Chance was unique—a throwback to how Brahman bulls appeared when they originally came to Texas from India in the 1920s as part of an effort to produce heartier, more disease- and drought-resistant stock. Since then, they've been remodeled, through cross- and selective breeding. Chance had all the traits Texas breeders had been working for decades to remove—pendulous folds of skin that sagged from his neck, a gigantic head, big horns, and a penis sheath that hung a foot and a half from his body and all but dragged on the ground.

"They've eliminated this skin in Brahmans now," Fisher says, using a pencil to point out the bull's privates in a photo. "Now, the sheath is way up here and the penis barely hangs down six or eight inches, as opposed to eighteen inches. The problem was, they would step on it. When a bull relaxes, this real soft tissue comes out of the sheath that protects the penis. So when they stand up, they stand on it and rip it and it gets swollen and infected in a pasture situation. Many a Brahman has died from that, or been unable to service cows for months and months and months."

Breeders managed to shorten the sheath, tighten the Brahman's saggy skin, and do away with the gigantic horns. Today's Brahmans have smaller heads and longer, more muscle-bound bodies. In a word, they are beefier. But at the same time, Fisher reflects, "they eliminated all the interesting things." In some respects, Chance was obsolete, like an Edsel, Fisher says; in others, he was exotic: "You have to go to India to see a Brahman that looked like that. Even the last few years we had Chance and took him to the rodeo, during Brahman week, all these old Brahman breeders would come up and say,

'What is that?' because even they didn't recognize that that's what they looked like in the 1920s and 1930s."

When not on the road, Chance generally hung around outside the Fishers' house—a sight that scared off more than a few delivery company employees. His favorite spot was under a tree, where Fisher's wife and business partner, Sandra, could see him from the kitchen.

"If I'd have let him, he would have come in the front door," Sandra Fisher says. "He would come walking up here and stand there and look in the kitchen window until I got the dog brush. I'd go out and start scratching him with it. Then he'd take two steps forward. That meant he wanted me to scratch his rump. Then he'd turn around so I could do the other side."

To hear the Fishers sit in their living room and reminisce about Chance, it's as if they were talking about a poodle. Chance was an outside pet, a money-making pet, but a pet all the same. And, they say proudly, he was probably the most famous, most photographed bull that ever lived. He met actress Ashley Judd and newsman Dan Rather, and Sonny Bono rode him at the Republican convention in Houston. Chance was a guest on NBC's *Late Night with David Letterman*. "Certain parts of this animal, it looks like he's shoplifting sporting goods," the talk show host quipped before climbing aboard.

In 1997, Chance had a major role in a disturbing and little-known movie *The Locusts*, starring Vince Vaughn, Ashley Judd, and Kate Capshaw as a cattle tycoon who, during the movie's climactic scene, castrates the bull. "They filmed it here in Texas," Sandra says. "It was a terrible movie, but he was the star."

Ralph's buzzard, Oscar, has a far more extensive list of movie credits, having appeared in seven. "If you see a buzzard on a dead person in a movie, it's probably Oscar," Fisher boasts. When animal rights groups questioned his keeping Oscar in captivity, Fisher built a nonprofit organization around him. That allowed him to keep the bird, whom he now uses in educational presentations.

While Oscar's Hollywood career soared, Chance was bringing home the bacon as well. For nearly fifteen years, Chance was part of Fisher's road show, going to rodeos (for picture-taking purposes), major sporting events, barbecues, conventions, and parades. Pulling out one of her many bulging scrapbooks, Sandra Fisher flipped to a photo of Ralph at a Galveston parade. He is standing atop Tumbleweed and Chance, one foot on each, holding the reins, as the procession moves down the street.

Tumbleweed's head and horns are now mounted on the wall of Fisher's living room. Other departed performers have been mounted as well, some of which are used in his presentations, including one full body mount of a steer on wheels.

"See, we keep things when they die around here," Sandra Fisher explains.

The living cast includes a dozen armadillos, an assortment of horses and ponies, two burros, Oscar the buzzard, and about ten Texas longhorn steers.

In June 1998, not long after Chance finished his movie, Fisher hauled him to Texas A&M to have a mole surgically removed. Eighty miles to the north, A&M's highly respected veterinary school was where ranchers turned for help with mysterious ailments among their stock, and, because the mole could have been a sign of a bigger problem, the Fishers wanted only the best for Chance, who was almost nineteen by then.

While Chance was there, a friend called Fisher about an article she had seen in the local paper. Texas A&M, it had been announced, was to serve as home for the Missyplicity Project, an anonymously funded quest to clone the world's first dog. The friend suggested to Fisher that, since Chance was already there, he could try to have him cloned.

"I said, 'What? What do you mean clone an animal?' She said, 'You know, make one from another.' I said, 'No, can't be.'"

Fisher did some research and made contact with the Missyplicity Project, only to be initially rejected by Texas A&M veterinarian Mark Westhusin. "We're researchers," Westhusin told him. "We're not in the business of cloning pets for people." Even though accepting funding to clone the funder's dog might appear as precisely that, to Westhusin it nevertheless was all about the research.

Fisher persisted. "We begged them. We said, 'Look, Chance is already there, and he's a great specimen for you because he's obsolete.'" When Chance came back home from A&M in early July, his mole removed and proven benign, the Fishers didn't know whether his cloning was being pursued or not. "At that point all we knew was, they had taken the mole off," Fisher says.

About two months later, in September 1998, Chance didn't return from pasture to his usual spot outside the kitchen window. When Fisher went out to look for him, he found him dead. Fisher walked back to the house for his camera.

"I was going to take some pictures. And then I thought, Well, he deserves more than that. So I skinned him. I skinned his whole body. It took me all day and it was raining. I'd skin awhile and cry awhile, I was just crying like a baby with my knife skinning. . . . I was a complete wreck by the time it was all over with.

"It was really a sad day. All the family had a good cry."

Fisher says that, to preserve the hide, it was important to skin

Chance in the first twenty-four hours after his death. "We were thinking of getting him mounted as a bucking bull and doing photos," Fisher explains, even though Chance wasn't a bull at the time of his death—he'd had his testicles removed two years earlier, because of disease. Since Chance had been shaved for his mole removal, the Fishers didn't pursue the full-body mount. Chance was buried on the ranch, and his pelt ended up in a box in their closet.

Chance had been dead four months when the Fishers received a call from Westhusin.

"I'm calling to see how Chance is doing," Westhusin said.

Only after Mrs. Fisher told him Chance had died did Westhusin explain that skin cells from Chance's mole had been cloned and that embryos, by then, were growing in seven different cows. A power outage at the university had caused the rest of Chance's cells to go bad—those being kept refrigerated and on reserve. Westhusin, not aware the original bull had died, was calling to get some additional cells from Chance for the purpose of comparing any cloned births with the original.

The Fishers were still able to accommodate the scientist, though. In addition to Chance's pelt being in the closet, other parts of him were in the family freezer. "When he died, I saved some of his tissue," Ralph Fisher says. "I'd cut off some skin and tissue and saved some blood and stuck it out back in the freezer." Fisher shipped the assortment of remains to A&M, where they were used to compare Chance's DNA with that of the embryos—most of which were being aborted along the way so that scientists could assess their progress.

"They said they had seven cows with seven viable clones developing," Fisher says. "But then they abort number one to see how it's developing. Then they abort another one, and another one. Then finally the geniuses decide, when there's only two left, 'Wait a minute, we better get a calf on the ground and then study the rest.'"

In August of 1999, A&M called the Fishers to tell them a birth was near.

"First it was just going to be a press release," Sandra Fisher recalls. "Then they called a day or two later and said they were going to do a press conference. I still wasn't going to go. But then they told us how big it was going to be—that there were 250 media outlets coming, from all over the world. It was more than just Texas's first clone, or Texas A&M's first clone, it was the fact that he had been a family pet."

At the press conference, the Fishers were given seats on the podium.

Veterinary school officials and scientists took turns at the microphone, drily explaining the cloning process, until one reporter asked Fisher to explain the science behind the miracle. Sandra Fisher thinks they were trying to make her husband sound like a fool, and it wasn't the first time, or last, during all the media coverage that the Fishers would get the feeling that they were being viewed as "country bumpkins."

To the surprise of the news media, and maybe the A&M scientists, Ralph Fisher, who majored in agriculture in college, and had done some research since, pretty much nailed it—though in a far more homespun way: "Yeah, it's real easy," he told the reporters. "You take an egg cell from a cow and you suck all the stuff out of it, then you put the donor's DNA in the egg and then it divides and you put it in a surrogate. Then seven of those cells became embryos."

"Easy" might have been a mischaracterization, but Fisher had the basics down. The tissue removed from Chance in the biopsy was thinly sliced, then diced with a razor blade into one- to three-millimeter pieces, then transferred into twenty-five-millimeter flasks containing a medium of fetal bovine serum and penicillin. As those were cultured at 37 degrees centigrade, eggs were obtained from a slaughterhouse, cultured, and their cells enucleated. Chance's cells were then pushed into the egg cells with a tiny glass pipette. Electric pulses were then delivered via a BTX Electrocell Manipulator 200. Dozens of other procedures, solutions, tests, and monitoring techniques were involved, all of which led to the implantation of fifty-eight embryos into twenty-three surrogate cows. Eleven fetuses reached the point of having heartbeats. But between miscarriages and abortions performed to assess their development, only three of those remained viable more than ninety days into pregnancy.

Only one was born—a full 290 days after Chance's DNA met the oocyte—a gangly white calf, the world's first cloned bull, delivered by cesarean section. He weighed eighty-eight pounds. He was named Second Chance.

He had his problems—lung dysmaturity, pulmonary hypertension, an iron deficiency, a yeast infection, and low leukocyte levels. For his first five days, he was fed with tubes. At seven days, he was diagnosed with Type 1 diabetes, which would require insulin to be administered for two months.

But to Ralph Fisher, he was pretty near perfection—a mirror image, he figured (though he never saw the original when he was a calf), of his beloved Chance. In interviews, Fisher repeatedly spoke of the cloned bull as the second coming of the first—and the news media, intrigued with the angle of a man reuniting with his dead pet bull, gobbled it up.

"In Texas tonight," one TV news report began, "a man, his wife, and their bull have been reunited." Ralph then appears on camera: "I guess he's been reincarnated."

The scientists at Texas A&M blanched at such references. They privately urged Fisher to refer to Second Chance as a "genetic twin" of the first bull. Fisher never did. "We really felt like we got him back. . . . The scientists kept saying, 'No, he's not the same animal; he's got the same genetics, but he's not the same animal.' But we felt he was. We just felt so blessed with this gift."

"A clone is the same as a genetic identical twin," says Westhusin—and nothing more. Its behavior and personality won't necessarily be the same as the original. But as much as he tried to persuade Fisher of that, he says, Fisher never quite bought it. "People get attached to their animals, and they want to sometimes see more than is there, and they do see more than is maybe really there."

Once he was brought home to the ranch, Second Chance started behaving in ways eerily similar to the original, the Fishers say.

"The first day he was here, he went over and lay down in the exact same spot that Chance always lay in," Sandra Fisher says. "Now that was spooky."

He had the same mannerisms. He licked Fisher's boots, just like the first Chance. He even ate the same way. Instead of keeping his head in the feed bucket, as most bulls do, he'd take a mouthful, pull his head out, raise it up, and chew.

"I'd never seen another animal do that," Fisher says. "I'm thinking he has the same instincts, he's going to make the same choices. I mean why would he even be eating the same way?"

The Fishers paid nothing to have Chance cloned; instead, all the expenses were covered by the billionaire who, though his identity had been kept secret at first, was known by then to be John Sperling, the founder of the University of Phoenix, who was said to be funding the Texas A&M cloning research in an attempt to get his dog duplicated.

The Fishers signed no agreements, turned over no rights, and had not been asked for anything in return for the cloning. So they were surprised when Sperling's front man, Lou Hawthorne, dropped by one day. They say Hawthorne wanted Second Chance to be the poster child for his newly established company—one that would bank the tissue of pets whose owners were interested in someday cloning them. Hawthorne saw Second Chance as an achievement worth touting further, and since Sperling's money had made his birth possible, he felt it was within his rights to ask. He wanted the Fishers to bring the bull to Chicago for an exhibit.

The Fishers, though their business revolves around putting animals on display, took some minor offense at Hawthorne's suggestion. "He's not a research project," Sandra objected, though that's exactly what he was. They insisted on being paid, and when cash wasn't offered, they declined to make the trip. "Maybe they felt as if we owed them," Ralph says.

As the bull's first birthday neared, the Fishers decided to throw a party for Second Chance—cooking up some barbecue for a couple dozen friends. But once Hawthorne and A&M got involved, it turned into a party for 350, many of them members of the news media. The Fishers made Second Chance a cake for the occasion, out of cattle feed, topped off with a couple of sparklers, which Second Chance immediately tried to eat.

After that, the Fishers would continue to throw Second Chance a birthday party each year, with smaller guest lists, and Second Chance continued to behave much like his predecessor—calm, tame, and predictable. On his fourth birthday, though, something snapped.

The guests had sung "Happy Birthday" to the cloned bull, and everybody had gotten their photos taken with him. Second Chance had finished off his birthday cake—a big tub packed with sweet feed (the Fishers had switched from sparklers to candles). The sun was going down, and it was time to get him back to the barn.

"I put one finger in his nose ring, and I had his lead rope in the other hand. We took a few steps and boom, he caught me with his horn and slammed me down. He just picked me up and slammed me down," Fisher recalls. "Then he proceeded to get on top of me with mostly his forehead and his horns. I said, 'Oh, no, this can't be happening.' All I could think of was, 'Why are you doing this? Why?' We just had so much faith in him."

Fisher spent the night in the hospital and was treated for a dislocated shoulder.

"It kind of put a damper on the party," Fisher recounts. It was the first time he'd seen true aggression in the animal.

The next day, Fisher looked at the ground at the scene of the attack, and saw several deep holes in the yard, inches from where he'd been lying, left by Second Chance's horns. He also took a closer look at the Texas-size belt buckle he'd been wearing the day before. Next to the inscription TEXAS FINALS RODEO CLOWN AND BARREL MAN, 1972, was a visible dent, caused by Second Chance's horns. Had the horn hit his stomach rather than the buckle, "he probably would have gutted me," Fisher says.

Even so, Fisher looked at Second Chance's outburst as a fluke. He reviewed the incident in his mind, questioning whether he might have done something to bring it on. Second Chance, Fisher was still convinced, was going to be every bit as tame and gentle as the original, because, as he saw it, he was the original, returned to life. Maybe, he reasoned, Second Chance was just going through a phase, and by the time he reached seven—the age of the original Chance when Fisher got him—he'd be just as calm and predictable as the original.

When friends and neighbors heard about the attack, they asked if Second Chance had been sent to the auction barn yet, assuming that the cloned bull would be sold for meat after the incident. The Fishers wouldn't dream of it. They found the suggestion insulting. Second Chance was a family pet, and he just needed some time.

"Everybody just assumed that since he hurt Ralph we would sell him," Sandra Fisher says. "We'd tell them, 'No, he's Texas's first clone.'

"You don't just get rid of a clone."

7.

The Raël World

■ ■ ■

When I die, I'm leaving my body to science fiction.

—*Steven Wright*

UFOLAND
VALCOURT, QUEBEC
2000

Lou Hawthorne wasn't the first to market the promise of pet cloning. At least three other companies beat him to the punch—two of which were based on earth.

Hawthorne was still seeking out his Missyplicity research team when Clonapet was formed, a branch of a larger company that—formed by a religious sect, incorporated in the Bahamas, and headquartered in a UFO theme park in Canada—announced its birth in a press conference at the Flamingo Hilton in Las Vegas.

Clonapet was a division of Clonaid, established by Claude Vorilhon, a Frenchman who changed his name to "Raël" after a religious experience in the 1970s. In March 1997, just one month after scientists in Scotland announced the birth of Dolly the sheep, the self-proclaimed prophet held a press conference revealing that his new company, Valiant Ventures Ltd., would soon be offering cloning to the public.

The goal of Clonaid—never officially incorporated in North America as a business—was clear from the start: cloning humans.

But to finance that research, Raël's followers, known as Raëlians, said they'd start off with animals. For one thing, it would be good practice. For another, they pointed out, "no legislation restricts animal cloning." Actually, no laws in the United States specifically prohibited cloning human beings, either, and that was one reason Clonaid was able to operate unfettered—to the extent it actually operated—until August 2001, when its plan to clone a West Virginia state legislator's dead infant son was exposed.

Clonaid began its website in 1997, and soon thereafter announced that its first subsidiary—Clonapet—would soon be open for business: "Clonaid, the human cloning company in the process of being structured, will soon offer a new service: the cloning of pets to well-off individuals who wish to see their lost pet be brought back to life. This service will also be offered to the owners of racing horses, a very promising market, given the outrageous prices paid for champions."

If the concept sounded out of this world, that might have been because, according to the book of Raël (and there are several of them), that's where it originated.

Less than two weeks before Christmas, on December 13, 1973, Claude Vorilhon was, by his account, driving through the French countryside when, on an impulse, he pulled over at Puy de Lassolas, an inactive volcano near Clermont-Ferrand, the capital of the

Auvergne region. He had hiked up the mountain and entered the crater when, he later recounted, a small silver spacecraft appeared. It landed, and a door dropped open. The three-foot-tall space alien who stepped out sported a beard and spoke French, and his skin had a greenish tinge—"a bit like someone with liver trouble," Vorilhon told London's *Sunday Telegraph* in a 2003 interview.

The son of an unmarried fifteen-year-old farm girl—or so he thought—Vorilhon had grown up living mostly with his grandmother and an aunt. At fifteen, he ran away from boarding school and hitchhiked to Paris, spending three years playing music on the streets until the director of a national radio program signed him up. Vorilhon took a new last name, Celler, and in 1966 recorded several singles, including "Monsieur Votre Femme Me Trompe" ("Mister, Your Wife Is Cheating on Me"), "Madam' Pipi" ("Mrs. Toilet Attendant"), and his lone hit, "La Miel et la Cannelle" ("Honey and Cinnamon").

A one-hit wonder by age eighteen, Vorilhon—having lost his luster as a pop star, and his wealthy and well-connected sponsor to suicide—set his sights on becoming a race-car driver and journalist, using the money from his singing career to start a racing magazine called *Auto Pop*. Through it, he gained entry to the racing world, and he drove professionally for three years—until November 1973, when the French prime minister instituted a moratorium on auto racing because of the gasoline crisis.

It was just a few weeks later that Vorilhon, twenty-six at the time and between jobs, had his alleged close encounter.

Though he was married, with the first of two children on the way, Vorilhon didn't hesitate to board the spacecraft when invited. The alien explained he had come from another galaxy—the planet Elohim—to meet with Vorilhon. Why him, and much more, became clear in the long talks that Vorilhon says ensued. He returned to the spaceship with his Bible, as instructed, for six straight days.

The alien spoke of the dangers posed to earth by war and nuclear weapons, and the need for earthlings of different nations to come together. God did not exist, the alien told Vorilhon, and evolution was a myth as well. All the life-forms on earth, from Adam and Eve on, were, in fact, clones. The Garden of Eden? It was actually a large temporary laboratory, set up on earth by maverick Eloha to surreptitiously pursue cloning, which was illegal at the time on their home planet. Noah's Ark? It was really a spaceship that carried DNA to resurrect animals through cloning.

With his newfound knowledge, Vorilhon began making arrangements for a large conference in Paris, where he announced the founding of the Raëlian Movement in 1974, and embarked on writing a book about his experience with the extraterrestrials. He sought out TV and radio appearances, and began publishing the Raëlian newsletter, *Apocalypse.* The next year, the aliens checked in on him again. On October 7, 1975, he says, he was taken to the planet Elohim. There he met Yahweh, the president of the planet, and dined with Jesus, Mohammed, Buddha, and Joseph Smith, founder of the Mormons—all of whom, it was explained, had previously been sent to earth by Elohim to do what they could in terms of guiding humanity.

Apparently, even with that guidance, the Eloha's intergalactic experiment to populate a planet with clones wasn't going that well, as evidenced by incidents such as Hiroshima and Nagasaki, and our continuing tendency to go to war with each other.

During his twenty-four hours on Elohim (a Hebrew word meaning "those who come from the light"), Vorilhon says, he ate well, spent the night making love to six female biological robots in a hot tub, and was made privy to the techniques of "sensual meditation." He claims to have viewed a clone of himself in a vat, and watched as a cell from his own forehead was placed into it, causing it to grow, in a matter of seconds, into an adult. And he learned that Jesus was his brother, by blood. Vorilhon was then returned to France with the knowledge that he was the latest in a series of messiahs Elohim has sent to earth—and that he was to be the last.

Back on earth, Raël divorced his wife and left France, choosing Quebec for his religion's headquarters. He continued writing books and gathering followers, among whom he encouraged kindness, inquisitiveness, and sexual freedom, all three being tenets of his newfound religion. Its membership—once limited to a core group of mostly science fiction junkies—swelled to the thousands, many lured in by the religion's promise of clean and peaceful living and all the sex you could possibly want. With its offbeat and futuristic beliefs and the occasional publicity stunt—such as handing out ten thousand free condoms to Catholic high school students in 1992—the religion received widespread press coverage.

To further spread the word, the Raëlians built UFOland, a museum and interpretive center in Valcourt, seventy-seven miles east of Montreal—the largest building in North America made entirely from hay and fiberglass. It opened in 1997, but was closed to the public four years later. Yet unbuilt—and, along with cloning, their biggest mission—is the embassy they plan to construct on earth for the expected return of their Elohim forefathers.

Raëlians put no stock in the idea that our souls live on when our bodies die—or that "souls" exist at all. One's only hope for eternal life, Raël preaches, is through re-creating the body from DNA, thereby giving one's "being" (not to be confused with "soul") a new, genetically identical, fresh-off-the-assembly-line receptacle to pass into upon the expiration of its physical host.

Initially, Raëlians believed that those among them who deserved it would resurface on Elohim as clones when their life on earth was over. Just as some religions encourage people to live their lives in a manner showing they are fit for heaven, Raël urged his followers to prove themselves cloneworthy. Each newly initiated Raëlian is encouraged to sign a contract with a Quebec mortician—for their convenience, there's a registration booth at Raëlian ceremonies—paying $500 and signing paperwork that gives the mortician permission to remove, upon their death, one square centimeter of their frontal bone (the large bone that makes up the forehead). The chunks are put on ice and shipped to Geneva, where they are kept in a refrigerated underground vault, Dawson University professor Susan J. Palmer explains in her book on the Raëlians, *Aliens Adored*.

Originally, it was said that Elohim spaceships would pick the specimens up and take them back to the mother planet for cloning. There, the technology was said to exist that allowed one's memories and personalities (already stored on Elohim's supercomputers) to be transferred to the clones. A "special machine" there, Raël says, is capable of uploading all of our information—from childhood memories to what we like on our pizza—into our newly duplicated shells.

From the religion's inception until 1997, cloning was something that, if you played your cards right, would happen to you—on a planet far, far away. But in 1997, with the birth of Dolly, it made the transition, for the Raëlians, from just reward to do-it-yourself project.

At the helm of Clonaid, Raël placed French biochemist Brigitte Boisselier. She was a vivacious and outspoken former deputy director of research at the Air Liquid Group, a French producer of industrial and medical gases, who had been fired for going public as a proponent of human cloning and as a Raëlian.

At one point, Clonapet claimed to have DNA from fifteen cats and dogs stored in special freezers. But in 1999, all mention of pet cloning disappeared from the Clonaid website, falling by the wayside amid newer services Clonaid was touting. Those included Ovulaid, which for $5,000 promised a catalog of human egg donors from which customers could pick the provider of unfertilized human eggs. It offered "a personal visit with prospective egg suppliers to better exercise your right to choose the appearance of your future baby." For $50,000, Insuraclone, another Clonaid service, allowed subscribers to send their DNA in for storage, then either use it to create a clone or reclaim it for use in stem-cell therapy in the event of illness or organ failure.

By the year 2000, Clonaid seemed to be stepping back from pet cloning—if it ever really stepped into it.

For their human-cloning service, though, the Raëlians were getting some bites, including one from Mark Hunt, a West Virginia lawyer who, amid a campaign for the U.S. House of Representatives, quietly arranged to pay Clonaid to produce a clone of his dead son.

Hunt, the eldest of five children and a graduate of Stonewall Jackson High School, was by then a forty-year-old lawyer who had served four years in the state legislature. In August 2000, through the Clonaid website, he made contact with Boisselier—at the time, a visiting professor of chemistry at New York's Hamilton College.

In September 1999, Hunt and his wife, Tracy, had lost their son, Andrew, who died after surgery to correct a congenital heart defect. He was ten months old. Hunt, who won a large settlement in a lawsuit against the hospital where the surgery was performed, had Andrew's cells frozen in hopes of someday cloning him.

Through an unregistered company called Bioserv, Hunt provided $500,000 to Clonaid for its cloning research, and he footed the $320-a-month rent on what would serve as its laboratory—an empty high school science lab in Nitro, West Virginia.

On the distant outskirts of Charleston, in an area along the Kanawha River known locally as "Chemical Valley," Nitro was founded and named by the federal government, which during World War I built an $80 million gunpowder and explosives manufacturing complex there. Just as the facility came on line—after only one outgoing shipment—the war ended. Nitro went on to become a center for chemical companies.

Given their intention to keep their facility's whereabouts secret, the location might not have been the wisest choice of venues for a covert laboratory trying to re-create a human. Even though it was off the beaten path in rural West Virginia, just southeast of a town named Scary, Clonaid's lab shared the building—a 1950s-era high school, reopened as a community center/business complex—with tenants that included a local police station, plumbing and roofing companies, a day care center, and a senior citizens' club.

In the same complex where toddlers finger-painted and seniors played bingo, Boisselier tinkered with cow eggs and, from most indications, nothing more. Her lab was in Room 201 of the old school. It was equipped with a telephone, computer, microscope, centrifuge, micromanipulator, and an incubator—at least some of the equipment used to create clones. Posters of cells hung on the wall. The operation was staffed by Boisselier, a geneticist, a biochemist, and a gynecologist affiliated with an in vitro fertilization clinic.

By that time, Boisselier had become an outspoken proponent of human cloning—so outspoken that she and Clonaid had drawn the attention of lawmakers in Washington. In September 2000, Raël—though not identifying Hunt—had announced in Montreal that an American couple was funding Clonaid's efforts in hopes of having a clone produced of their dead child.

In March 2001, when called to testify before a U.S. House of Representatives subcommittee working on a bill to outlaw human cloning, Boisselier, speaking against the bill, read a letter from the anonymous client—a person she described as a father who had lost his ten-month-old child: "I am a successful attorney, a former state legislator, a current elected official, a husband, a son, a brother, but most importantly, I am a father," the letter read. "We didn't know what to do and couldn't accept that it was over for our child, and for the first time in human history, I/we didn't accept death as the end. Not since our Lord and Savior, Jesus Christ, spoke to Lazarus and told him to come forth from the grave has a human been able to bridge the great gulf of death. . . .

"I hoped and prayed my son would be the first; I decided then and there that I would never give up on my child. I would never stop until I could give his DNA—his genetic makeup—a chance. I knew that we only had one chance: human cloning. To create a healthy duplicate, a twin of our son. I set out to make it happen."

Boisselier's remarks before the subcommittee—she all but guaranteed a human clone birth within the year—brought her under more scrutiny from federal authorities. When, later that same spring, Food and Drug Administration officials learned of her lab's location—via the school building property manager, who saw Boisselier on TV, realized what she was doing, and called police—it sent investigators to pay a visit. They found no human eggs, only those of cows. They seized the lab's equipment and, in the aftermath, hammered out a deal: in exchange for ceasing all cloning activity, Boisselier would not be charged. While there was (and is) no specific federal law against cloning a human, authorities had considered charging her with violating U.S. drug laws. Also as part of the deal, Hunt's involvement in cloning and his financial support of the lab would remain a secret.

Hunt, a Democrat, had originally entered the 2000 race for U.S. Congress, but later dropped out and ran instead for West Virginia State Senate. He lost that race—even though his failed quest to clone Andrew stayed secret until August 2001, when a journalist from *The Sunday Times* of London showed up in Nitro.

Between Boisselier's public comments and Internet search results, reporter Joe Lauria was able to put the pieces together, locate the lab, and pinpoint Hunt as the anonymous donor. The story came out days after the U.S. House of Representatives voted to ban all forms of human cloning, a measure that didn't pass the Senate then and hasn't since.

Even after the secret agreement became public, Boisselier continued to tout cloning and seek clients (at $200,000 a shot), and vowed to press on with her work—if not in Nitro, then in a new secret location. She told Lauria that fifty women had come forward to bear the Andrew clone, including her own twenty-two-year-old daughter. In addition to Hunt, she said in a news release, two thousand more potential clients had contacted her, interested in creating clones of themselves or loved ones.

Hunt, who was never a member of the religion, backed away from the Raëlians, at least publicly, announcing he had severed his ties with Clonaid and Boisselier, whom he termed a "press hog." Hearing that she planned to continue the quest to clone a human, he went so far as to change the locks on the schoolhouse lab door. Hunt would go on—after a four-year hiatus from politics—to get elected again to the West Virginia House of Delegates in 2004, where he still serves.

8.

"I Love Perfect Things"

■ ■ ■

> Inside of me there are two dogs. One of the dogs is mean and evil. The other dog is good. The mean dog fights the good dog all the time. Which dog wins? The one I feed the most.
>
> —*Anonymous*

COLLEGE STATION, TEXAS
February 2000

Two years into their research, the scientists at Texas A&M—despite their success with cloning a bull on John Sperling's dime, and despite continual refinements to the process of renucleating canine eggs with the DNA of Missy—had yet to duplicate the dog that was still being portrayed as the billionaire's beloved mutt.

Every month, researchers would select bitches in heat from their stock of laboratory dogs—mostly beagles—anesthetize them, surgically remove their eggs, and transport the cells across campus to another lab, where Missy's DNA would be inserted and the combined cells would be fused with a jolt of electricity. The newly formed cells would be toted back to the original building to be implanted into still more laboratory dogs—none of whose ensuing pregnancies would, by 2000, get to the point of having a heartbeat.

Every other month, Lou Hawthorne would stop in. Usually, he'd bring his video camera, or a videographer, to shoot some footage for the documentary he planned to produce and sell once the Missy mission was accomplished. On rare occasions—to the chagrin of Westhusin and other scientists—he'd invite a reporter from his news media "A-list" to come along for a glimpse. In 2000, it was Charles Graeber, from *Wired*, who tagged along and would later recount this scene:

The team of scientists had removed and isolated the egg cells of a dog in estrus, sucked out their nuclei, and re-enucleated them with Missy's DNA. The next step was to fuse them, allowing them to break through their individual membranes and form one cell—a job for the BTX Electro Cell Manipulator.

Taeyoung Shin, a former veterinary student at Seoul National University in South Korea, who was pursuing his research at Texas A&M, was at the controls of the device, which was about the size of a car battery, with a series of dials and switches to set voltage and timing, and two protruding wires that attach with alligator clips to the sides of the dish containing the cell. When Hawthorne zoomed in on the machine with his camera, Shin held out his hand, blocking the camera from capturing the settings he had keyed in.

Shin told Hawthorne the settings were secret—"Really, really, it's secret, you cannot film"—but Hawthorne only stepped closer, continuing to film and at the same time letting the scientist know who was boss.

"Well," Hawthorne said, "I'm paying for it."

As the conduit through which the funding for the Missyplicity

Project flowed, Hawthorne had the right, in his view, not just to film what he darn well pleased, but to reap the profits from the technology, once conquered, as well.

On Valentine's Day, 2000, he took a big step in that direction, with the announcement of the establishment of Genetic Savings & Clone, a company that would allow pet owners to bank the DNA of their dogs in hopes that someday—soon, it promised—it could be used to clone them.

By then, at least two other companies had positioned themselves to cash in on pet cloning—not counting Clonaid, the much-doubted Raëlian company that claimed to be working on human and animal cloning.

PerPETuate was founded in 1998 by veterinarian Heather J. Bessoff and agricultural consultant Ron D. Gillespie to store the cells of pets for owners who thought they might one day want to clone their animals. It initially operated out of Bessoff's basement. The company provided collection kits that pet owners could take to their veterinarians, who, through a biopsy, would remove a small sample of tissue and overnight it to Cyagra, the livestock-cloning company with which PerPETuate was affiliated.

Lazaron BioTechnologies, formed that same year by two Louisiana State University scientists, Richard Denniston and Brett Reggio, was also banking the cells of people's pets by 2000. "Does your loved animal's singular genetic character have to die?" the Baton Rouge–based company named after Lazarus, the biblical figure whom Jesus brought back to life, asked on its website. "No. It can live on."

Hawthorne made the same claim on the Genetics Savings & Clone website:

"Chances are you're a lucky pet owner," read the introduction on the Genetic Savings & Clone website. "Your dog or cat is exceptionally compatible with you, possessing the right mix of temperament, intelligence, and good looks for your taste and lifestyle. If you're like most pet owners, your animal is a spayed mixed breed, truly one-of-a-kind. Get ready for lightning to strike again. . . . Thanks to cloning, soon you won't have to rely on luck to have another pet very much like yours. After five years of research, we believe we're less than a year away from offering commercial cloning of dogs and cats. The first step is to preserve your pet's DNA in our gene bank. It's simple, it's safe, it's forever."

All three companies were banking pets' DNA by 2000, but only Genetic Savings & Clone was, through Texas A&M, actively pursuing the cloning of dog.

The relationship between the company and the university was an awkward one, especially considering the initial involvement of some of the university scientists with the business side. Genetic Savings & Clone termed Westhusin a "founding partner" in its initial press releases, but all the scientists involved with Missyplicity—except for Shin, who went on to join the staff of GS&C—say they have severed any links to the business. Some of the scientists were bothered, too, by Hawthorne's marketing claims, and the question of whether they were taking advantage of grieving pet owners. With Sperling's continued funding, though, those concerns stayed mostly beneath the surface.

On top of the initial $2.3 million he gave Texas A&M in 1998, Sperling supplied the university with $2 million more in 2000 for its cloning research, and he sank an additional $10 million that year into the start-up of Genetic Savings & Clone.

Hawthorne, in his business plan at the time, didn't see gene banking as a big money-maker. It required only a $1,000 setup fee and $100 a year to store a pet's tissue. That alone could make millions for

Genetic Savings & Clone, but it was small change compared with the profits being the sole worldwide provider of dog cloning could produce. Once successful in cloning Missy, Hawthorne predicted, the company would be cloning dogs for the public within six months, at a cost of $250,000 each—a price that would drop as the process was improved and refined.

Estimating there were 50 million dogs in American homes at the time, and factoring in their short life spans, he calculated 5 million dog deaths a year. With his company serving as the guardian of their DNA, and holding the patent rights to the technology used in creating Dolly the sheep, he would, once dog cloning was achieved, have a captive audience in more ways than one.

Despite their not having produced a viable pregnancy, the scientists and Hawthorne knew it was only a matter of time: enough trial and error would eventually lead to a cloned dog. The bigger challenge lay in selling the public on pet cloning and avoiding the wrath of animal welfare organizations.

To counter the criticism he knew would be coming, Hawthorne established the "Missyplicity Code of Bioethics," which spelled out a new and higher standard of care for laboratory dogs, who, in the United States, are generally dogs bred to serve that purpose. They commonly live out their lives in cages and often are disposed of once their services are no longer required.

For the Missyplicity Project, Hawthorne ensured the animals received care and treatment that went far beyond prevailing laboratory standards, including giving the dogs names, as opposed to numbers, and finding adoptive homes for them when they completed their term of service. The project also limited to eight months the time that egg-donor dogs and surrogate dogs were kept in the rotation.

An A&M graduate student, Jessica Harrison, was placed in charge of tending to the dogs, caring for them in a bungalow a mile from campus, socializing them after they were used in experiments, potty-training them, and finding adoptive homes for them when their services in the laboratory were no longer required.

"We slowly exposed them to all the things they'd be exposed to in a family home—like TVs, mirrors, grass, trees, flowers, birds, and bees," she says. "These dogs had never seen any of that. You put them down on the grass, and they're like, 'What's this?' It was kind of overwhelming. You get used to it, but at first it's like, these are dogs, how can they not know these things?"

Harrison would have students come in and help socialize the laboratory dogs, who—for reasons of nature, nurture, or simple lack of opportunity—are not inclined to play. "What they teach them is to be still," Harrison says. "As puppies, they teach them to just freeze when a person messes with them. We had to kind of undo that and say, 'No, we want you to move around and be excited.'"

The code of ethics—and finding adoptive homes for every dog used in the research—didn't totally silence critics, but it helped.

There was never a single demonstration by animal welfare organizations in connection with the quest to reproduce dogs by cloning—perhaps a testament to the public relations savvy of Hawthorne, a man who is something of a reproductive marvel himself.

He was born to a woman who thought she had lost any chance to have a baby. The culprit, she suspected, was technology of another sort: nuclear and chemical weapons. Joan Hawthorne's first pregnancy resulted in a stillborn infant with a malformed head. Multiple miscarriages followed. By her mid-thirties, she believed—with what would turn out to be ample reason—that her eggs, and possibly more, had been damaged as a result of working in nuclear and chemical weapons laboratories in the late 1940s and early 1950s. She'd tried having a baby with one husband, then another. She'd even tried early forms of artificial insemination, without success.

But then, in 1958, she had a daughter, and in 1960, a son.

"Our son arrives on his due date and sucks eagerly," Joan would later write of her son's birth in a New York hospital. "There is lots of time to note and inspect his slightly mismatching ears, to kiss them and hug him closer when the gray fear threatens to envelop me. Ten fingers, ten toes, beautiful eyes, an early smiling mouth, why should ears match? One is elongated; the other has a small dip, or flop-over. . . . He never cries. My mother wonders if he is all right—shouldn't he cry?"

As a toddler, Lou bounced back and forth between his mother and father—until Joan was able to permanently wrest her children away from him. She raised them mostly by herself, and mostly on the lam. She defied court orders, and manufactured a new identity—opting to take the last name of an American novelist. It made for a bumpy childhood, with stops in Italy, Oregon, Arizona, California, and, more than a month, when Lou was only four, in a juvenile facility.

Joan Hawthorne would recount the troubled years in three published memoirs, written under her pen name Candida Lawrence, the last name also borrowed from an American novelist. Two of the memoirs deal with her break from an abusive husband, a professor of medieval art, and her life on the run. The third recounts her quest to determine if her employment in the weapons industry in the late 1940s and early 1950s was responsible for her fertility problems and, later, breast cancer.

"Yes, growing up in hiding and all that is factual," Lou Hawthorne said. "I think it was psychological warfare with my mom. My father wanted to hurt her in the worst possible way, so he took her children. That will do it. But what he wasn't counting on was that she was going to take the war to the next level. He won all the battles, but she won the war. He looked for us for six or seven years after we disappeared. We know that from mutual friends."

In an interview at the dog park near his home in Marin County, California, Lou said his father died in 1996—at a time when he was thinking about reestablishing contact with the man he hadn't seen since age five. Sitting on a park bench, he shrugged off condolences: "That's all right. He was a son of a bitch for taking two small children from their mom. You get what you deserve."

While the courts repeatedly awarded custody of Lou and his sister to their father, their mother repeatedly fled with them, the third time for good. She settled in California, borrowed a friend's college transcripts, and got a job as a schoolteacher. "It all came at quite a price," Lou said. "She had to re-create her identity, and forgo all her straight-A transcripts from Berkeley and her degrees. She swapped transcripts with somebody with C's and D's from some Podunk school. She had to pretty much start over. . . . She wasn't able to make any investments along the way . . . because of the illegality of her records. . . . So we lived hand-to-mouth. It was an intellectually privileged but economically lower-class upbringing, you could say."

In her memoirs, Joan recounts the instability of those years, and its impact on her children. Lou, who is called "Tony" in the books, was a curious kid. As a toddler, he would shake his crib until it would roll across the room, hurtle himself out of it, and crash to the floor, then come to her room, where she pretended to be sleeping, and pry her eyelids open. She recalls a morning that he crawled outside with a small box of Rice Krispies and tried to feed one to a worm, and the notation from a Montessori school assessment that "the three-year-old still carries with him a heavy burden of chaos." And she remembers the fearful day when "Tony" fainted and was taken to the hospital: "He submitted to electrodes and blood extraction with a great deal of interest." The tests showed all was normal. The youngster did have a lisp, though, and a curious habit of repeating the ends of his sentences: "He says, 'I think electhrithity can get out of the walls and go everywhere,' looks down at the floor and whispers, 'walls and go everywhere.' . . . 'Tony, you are repeating your words in a whisper, why are you doing that?' "

In two of the books, she recounts her affair with a married professor named Jack Davey—actually John Sperling, and how her husband, "David," a college professor and no exemplar of fidelity himself, found out about it. She paints her second husband—the father of her children—as a cruel and controlling man who behaved inappropriately with her daughter. Despite those claims, the courts would grant custody of the children to the father. In 1964, during a visit, she fled with the children to California, but authorities tracked her down and placed the children temporarily in a juvenile facility, where they would remain for more than a month, because their father was overseas. She would be allowed to spend thirty minutes visiting them twice a week.

In the summer of 1965, she absconded with them again, moving first to Tucson. After two weeks living in shabby motels, she took the children to California, where, in "a town by the sea," she rented a house, enrolled them in school—no easy feat without birth certificates—and found a job at a small private school.

"Tony" had occasional problems at his school, mostly for talking so much in class that other students missed the lesson. Once, he was kept after school to write "I will behave in class" on the blackboard, fifty times.

Their home was near Ventura, California, but only a select few knew that, among them John Sperling, who visited regularly, helped support the family financially, and bought them gifts for the holidays. In 1966, Sperling gave the boy a Sting-Ray bicycle for Christmas, according to the second memoir. "'Oh, whenever I think of my Sting-Ray, a light goes on in my head,'" mother quotes son as saying. "'It's perfect. I love perfect things!'"

"Tony" had some pets as a child—a bird, then guinea pigs, and two gopher snakes, one of which the boy took to wearing around his neck while riding his bike. Later, the family got a dog—one they believe to be a shepherd-coyote mix—and a cat, which, on its first day home, was bitten by the dog.

"He lay motionless for three days in a little kitty-coma," Lou Hawthorne would later write, "and when he finally opened his eyes and jumped up we all discovered . . . that he could now only turn left. . . . He would recognize one of us, run in our direction, then veer to the left like a train on invisible curving tracks. . . . We enjoyed Lefty's company for several weeks, until one day he simply disappeared. . . . We think he went exploring, and just ended up someplace where home was to the right and a strange and beckoning world lay to the left."

Hawthorne would graduate from high school and get a scholarship to Princeton University, where he majored in creative writing. After college, he worked as a software designer and cinematographer, finding work as an editor of documentary films. As a young professional, he went in some circles himself, and tended to veer toward those parts of the world that were strange and beckoning. He's a practitioner of martial arts, a student of Buddhism, an avid fan of the *Star Trek* movies.

In 1991, he became involved with Biosphere 2, a privately funded experiment outside Tucson and on the fringes, most have since concluded, of science. Four men and four women were to be sealed inside a glass-and-steel structure for two years, living independently and untouched by the outside world. A multimillion-dollar project privately funded by Edward P. Bass, whose family made a fortune in oil in Texas, the Biosphere would be self-supporting, much like an outpost on Mars—with its own ocean, savannah, and 3,800 species of plants and animals all sealed inside.

Hawthorne was hired to produce an educational documentary on Biosphere 2 by the University of Phoenix, Sperling's school, which, according to a *New York Times* article, had "close" but unspecified ties to the project. Not long after the Biosphere was sealed, one of the Biospherians left the premises to get medical treatment for a cut finger. When she returned, it was quietly and with a duffel bag full of new supplies. Word leaked out that the Biosphere wasn't as "sealed" as it had been made out to be, and among those doing the leaking was Hawthorne. Hawthorne spent several weeks there before the sealing, and said he learned that occupants had, contrary to public statements, packed in a full year's supply of food. They'd also rigged their computer software to allow the editing of data, he said.

When he joined the chorus of those raising questions about the project, Hawthorne was sued by both Space Biospheres Ventures and the University of Phoenix, which was apparently expecting the "documentary" to be a strictly positive account. The lawsuit charged that Hawthorne had misappropriated trade secrets and confidential information and breached his contract, and it requested that his notes and video footage be turned over. Hawthorne says the later-settled suit was prompted by his refusal to do a positive piece, and discussions he had with ABC-TV about co-producing a Biosphere documentary with the network.

Hawthorne continued to work on films after that and later launched a fund-raising effort to produce his own, *Hell's Buddhas*, a documentary about a group motorcycle trip across India. For more than a year, he worked to raise money for the production, but by the time he arrived in India, he still lacked enough funds to pull it off. Hawthorne informed investors of that, had "a good cry," and questioned his self-worth and identity: "Who are you now that your project is dead, now that you aren't doing what you told everyone in the world you were going to do, now that you owe all this money and have nothing to show for it?" he'd write in a continuing online journal about the project. He recalled complaining to his girlfriend: "I just wish there was something I could count on." She replied, "What about impermanence?"

While in India, he got word of his father's death. He called to check in on his mother from the foothills of the Himalayas. She told Lou his father had had a heart attack while lecturing at Yale. In another entry in his online journal, Hawthorne noted that he had only recently started thinking about getting in touch with his father. He had hoped, he wrote, "that we would meet again; that he would forgive me and my sister for not seeking him out, when we were children, or later, when it could have been easily done; that he might forgive . . . my mother for stealing us and hiding us all those years; that I might forgive him—as I've forgiven my mother—for turning my childhood into a battleground; and finally that he might share with me the kind of secrets that Robert Bly prattles on about, secrets men only impart to their sons, away from women, in quiet reverent voices, with the TV on 'mute.' Perhaps such secrets don't really exist. I wouldn't know. That's my point."

The man who came closest to filling that role for Hawthorne was John Sperling, whose relationship with Hawthorne's mother—though they aren't legally married or permanently cohabiting—has continued unofficially for more than forty years. Sperling has served as his role model and father figure, and as master to his apprentice.

"Lou's not really his son, but he's kind of his adopted son, in terms of the relationship they've had over the years," a onetime business acquaintance says of Hawthorne's relationship with Sperling. "But there's no doubt they love each other like father and son do; and for Lou, what John thought has always been very important to him. I think he wanted to impress him, like any son would want to impress his father."

Cloning Missy was Hawthorne's chance to make his hard-to-please mentor proud, and, along the way, maybe answer some of the questions he had long pondered about nature versus nurture. How much of what we become—or what a dog becomes—is shaped by genes, as opposed to environment and experiences? Are we our parents? Was he his father? Would Missy's clone like broccoli?

Whether Hawthorne found those answers or not, failure wasn't an option when it came to his new mission—cloning the world's first dog, and turning pet cloning into a multimillion-dollar business.

He had something to prove, both to himself and to Sperling.

9.

Positively Cloneworthy

■ ■ ■

For the pride of trace and trail was his, and, sick unto death, he could not bear that another dog should do his work.

—*Jack London,* The Call of the Wild

PROSPECT BAY, NOVA SCOTIA

September 11, 2001

Four months after his official retirement from the Halifax Police Department, Trakr had made the full transition to family pet, though, in truth, he'd been one all along. Cuddled, coddled, and doted on, the German shepherd, even after his crime-fighting days were over, still went nearly everywhere with his handler and partner, James Symington—including on vacation.

That's where Symington was in September 2001. On an extended sick leave from the department, he, his wife, and another couple were enjoying a few days in the quiet solitude of nearby Prospect Bay.

Trakr—no longer searching, rescuing, tracking criminals, or sniffing out stolen merchandise—was out of the limelight by then, enjoying the life of a normal dog. He was lying at the feet of his master when, on the morning of September 11, Symington flipped on the television in their cottage and learned that the world had been rocked.

Symington, along with his friend Joe Hall, a police officer visiting from California, watched in silence as the World Trade Center—first one tower, then the other—collapsed. There was no debate, and little discussion. Once they accepted the reality of it, grasped the magnitude of it, their first response, like that of many others across the globe, was to wonder what they could do to help.

"As I asked the question," Symington recounted, "the most obvious answer was right at my feet."

Symington and Hall loaded Trakr into the car and headed for New York City, a fifteen-hour drive. By eight a.m. on September 12, Trakr, one of the first dogs to arrive at Ground Zero, was navigating through the smoking ruins. Symington and Hall led Trakr through the devastation, climbing up and down the unstable piles of twisted steel, pulverized concrete, and shattered sheetrock in hopes that the dog—amid a thick scent of dust and death—might detect some life.

On one pile of rubble, Symington says, he did.

Trakr, though he didn't give a full-blown alert, according to witnesses, showed interest. He pawed and sniffed at the ground, then sat. To Symington and Hall, he seemed to have found something. "It was very chaotic, very hard to see, very treacherous, but for one quick brief moment, Trakr gave me an indication that someone was buried below the area that we were searching," Symington said. According to others who were present, the two men decided to get a second opinion, calling over another dog and its handler. The second dog gave a mild alert as well.

Symington said they were looking under the debris when a warning came that a nearby pile of rubble was about to fall. A firefighter hung his jacket at the spot that Trakr had pinpointed and Symington and Trakr moved on to search other areas.

For the rest of the day, they would find only human remains, some of the 19,858 pieces that, one year later, New York City's medical examiners were still trying to connect to the more than 2,800 who died. It was frustrating, sickening, and discouraging work, but Symington was buoyed when he heard—the next day—that a survivor, Genelle Guzman, had been found in the very spot where Trakr had alerted them.

"A couple of the firefighters we were searching with came by and congratulated us. They said, 'The area where you got the hit is where they pulled a lady out alive.' Subsequently, we talked to some of the rescue workers that were involved that physically pulled her out, and they confirmed and gave credit to Trakr as the dog that led them to the area where she was pulled from the debris."

Trakr was one of an estimated three hundred search-and-rescue dogs from across the country that responded to 9/11, pressed into duty at the World Trade Center, the Pentagon, and Fresh Kills, a landfill that served as a depot for the debris to be sifted through in search of human remains. Of those working at Ground Zero, some dogs were trained to find cadavers, some were trained to find live people, and a few were trained to find both. They worked twelve-hour shifts, many sustaining injuries in the piles of debris.

While search-and-rescue dogs are trained to see their work as a game, getting rewarded with a favorite toy when they make a find, the long hours, toxic conditions, and dismal surroundings took its toll on many of them. Some handlers of dogs trained to find live humans spoke of their dogs seeming depressed, and went so far as to get volunteers to hide in the debris so that their dogs could find them and get some positive reinforcement.

The reverse was even more common—instances of the dogs lifting the morale of rescue workers, and even comforting the families of victims. In addition to the job they were trained to do, many of the dogs served as unofficial therapy dogs, noted Lieutenant Dan Donadio, the head of the New York Police Department's canine unit, who supervised the efforts of search-and-rescue dogs in the aftermath of 9/11. Petting the search-and-rescue dogs, or even just looking at them, gave some at least a small sense of renewal.

Donadio has stated that search-and-rescue dogs made no live finds in the days after 9/11. And he told the Halifax *Daily News* in 2001 that Symington didn't report for duty with him. But, given that security was lax, and several visiting dog-and-handler teams did manage to get in, he doesn't rule out the possibility.

The chaos of the moment, the passage of time, disputed facts and conflicting news accounts have all blurred the history of the rescue of Genelle Guzman, a thirty-two-year-old immigrant from Trinidad who worked as a clerk for the Port Authority of New York.

She was on the sixty-fourth floor of the North Tower when she felt the building shake. As smoke began appearing on her floor, she and some coworkers rushed to the elevators, saw that they weren't working, and headed for the stairs. After descending fifty-one flights, Guzman, her feet throbbing, stopped to take off her high-heeled shoes. At 10:28 a.m., as she leaned over on the thirteenth floor landing of Stairway B, the ground fell out beneath her. She heard a deafening roar and, in total darkness, she felt herself falling.

When the roar subsided, she was lying on her right side, her hair trapped under a giant slab of concrete, her right leg and arm pinned under steel beams and more debris. In desperation, she jerked to free her head, ripping the cornrows from her scalp. Using her left hand, she tried to free herself, but it was useless.

The youngest of thirteen children, and a mother of two, who had moved to the United States three years earlier, Guzman began to pray, "Lord, just give me a second chance, and I promise I will do your will." Then she rested her head on the one soft spot she could find amid the rough concrete and sharp edges. Upon that cushion—it turned out to be the leg of a deceased firefighter—she fell asleep.

She didn't awaken until she heard the beeping sound of a construction vehicle. She heard people talking, and called out. Getting no response, she called out again. This time, someone answered: "Hello, is somebody there?"

"Yes, help me. My name is Genelle and I'm on the thirteenth floor." She stuck her hand out through a tiny crack that daylight was shining through. "Can you see my hand?" No one answered, and Guzman lost consciousness.

When she awoke, she could hear rescuers directly above her, and she called out again. She stretched her hand through the crack and, this time, someone grabbed it. It was about twelve-thirty p.m., and she had been trapped twenty-six hours.

"Genelle, don't worry, we're getting you out," the rescuer said.

"What's your name?" Guzman asked from beneath debris.

"Paul," the answer came back. "I won't let go of your hand until they get you out."

Hearing news of a survivor, more rescuers converged on the spot, with pry bars, saws, and a cutting torch. They managed to free Guzman's arm, then her legs. For twenty minutes, they pulled debris off her body until they were able to lift her out and place her in a stretcher. Applause broke out as the stretcher was passed hand-to-hand down the pile by a line of rescuers. She was only the fourth civilian to be rescued from the rubble, and she would be the last.

In the hospital, Guzman would ask about Paul, the firefighter who was the first to reach her, who kept talking to her and spent nearly half an hour holding her hand, often with both of his own. "Sometimes I wonder about Paul, the man who held my hand and calmed me when I thought I couldn't go on," Guzman would say years later in an interview with her pastor, Jim Cymbala of Brooklyn Tabernacle Church. Cymbala included Guzman's 9/11 account in his book *Break Through Prayer: The Secret of Receiving What You Need from God*.

No one seemed to know who Paul was. Paul never stepped forward—not then, and not since—but in the days following 9/11, others would.

Two volunteer firefighters from Massachusetts, Rick Cushman and Brian Buchanan, attached to the Pittsfield Emergency Management Agency, were quoted in several early news reports about Guzman's rescue, and they would be reunited with Guzman by CNN, more than three months after 9/11, in an interview moderated by correspondent Gary Tuchman.

TUCHMAN: She survived a choking avalanche of concrete and dust. Buried alive in total darkness, Genelle Guzman lay wedged in the rubble for twenty-seven hours, until rescuers finally heard her cries. They had rushed down to New York from Boston on September 11 to help with the rescue efforts. . . . Exactly a hundred days after the attack, we reunited Genelle Guzman with her two rescuers. . . .

CUSHMAN: The reason you were found was actually because they spotted a fireman's jacket, and the basic rules are firefighters take care of their own, so a firefighter went up to get him and that's how you were found.

BUCHANAN: Just as she got to me, she sort of opened her eyes and looked up and, you know, kind of asked me if she was out yet, and I said, you know, "You're just about there. You're good to go, you know, just hold on just a few more minutes and you'll be all right."

TUCHMAN: And, Genelle, do you remember saying that?

GUZMAN: Yes, I can remember saying that.

TUCHMAN: Do you remember that face?

GUZMAN: No, I can't remember the face.

BUCHANAN: I had less hair. . . .

TUCHMAN: Now Genelle Guzman can see her rescuers clearly. They are moved and amazed by her survival. . . .

When Guzman asked them about "Paul," the two men said they could remember no one by that name at the scene.

That Paul Somin, now a lieutenant in the New York Fire Department, never stepped forward didn't surprise those who know him. Still, seeing two volunteers from Massachusetts taking credit for the rescue on CNN infuriated some of his fellow members of Brooklyn Firehouse Rescue 2.

"I'm home one morning and I clicked on the TV and on CNN they were talking about reuniting Genelle with the people who found her," said Bob Galione, another member of FDNY's Rescue 2. "In walk these two guys, and I'm like, 'Who the hell are they?' Were those two guys maybe the last two guys at the end of the line of a hundred people as they passed her down? I guess that's possible, but they certainly and absolutely were not anywhere near where this was going on. . . . It amazes me that people are still trying to take advantage of what happened up there."

Galione said the reason rescuers were checking the area Guzman was pulled from was not because a firefighter's jacket was seen there, as some news accounts suggested, but because another survivor had been pulled from the same area earlier.

"Paul was with Billy Esposito," Galione said. "They were crawling up there into no-man's-land, and there was no dog in sight. . . . Paul was calling out, 'Anybody there? Anybody there?' and he thought he heard something, so he would dig a little, and call out, dig a little more, and call out. He reaches his hand in and the next thing he knows he has a hand holding on to him. It kind of freaked him out. . . ."

Galione said he and his fellow firefighters generally avoided the news media, and looked down upon any of their ranks who stepped into the spotlight. Glory-seeking was not the FDNY way; it ran contrary to their thinking that what they did, they did as a team. That firefighters were elevated to hero status as a group after 9/11 was fine, but being singled out was frowned upon, and guaranteed to lead to some razzing. When Billy Esposito was captured at Ground Zero in a now famous photograph of three firefighters raising an American flag—an image later used as model for artist Stan Watts's "To Lift a Nation" sculpture at the National Fallen Firefighters' Memorial—he was derided by fellow officers.

"We tortured him about it. 'You were setting up a flag while we're digging for missing people?' Billy still gets chewed on," said Galione, who retired in 2009 after thirty years with the department. "We had a policy of no pictures, no interviews. We took a few cameras away from people who were posing in front of the ruins, and we smashed them. I don't get it. It's not fucking Disneyland."

What is clear in the hazy aftermath of 9/11 is that the news media—seeking some much-needed uplift amid the carnage—may have been too quick to confer hero status. Some of those praised in press accounts would turn out to be, at best, only on the fringes of the rescue, if even that. As with Hurricane Katrina, four years later, the days and weeks after 9/11 tended to bring out the best and worst of humanity—from selfless feats of quiet heroism to shameless acts of self-promotion.

One of the most disturbing examples of the latter came from Raël, the pro-cloning religious leader in Canada. Raël issued a press release raising the possibility of using the victims' recovered body parts to clone everyone killed in 9/11. Just as the World Trade Center could be rebuilt, so too could all the humans, he said. All the victims could be cloned and returned to their families, and the perpetrators who lost their lives in the attack could be cloned, put on trial, and executed. The announcement merited only the briefest of mentions in the news media.

Receiving far more publicity, and a goodly amount of donations as a result, was Scott Shields, who had shown up at Ground Zero with his search-and-rescue dog Bear, an aging golden retriever. Shields would go on, after 9/11, to recount his dog's heroics in public appearances—how he and Bear were the first to arrive at the scene, and how they made more "live finds" than any other dog at Ground Zero.

In truth, Shields and Bear found no one, and they had been asked to leave the site shortly after they arrived—because of their lack of credentials.

Symington, by contrast—though he responded without his police department's authorization, though he never registered with emergency coordinators at the site—put in long and hard hours in the days after 9/11, leaving Trakr sick and exhausted.

Trakr's role in finding Guzman would never become a big story in the mainstream media, though Symington, while still at the scene, did recount it for a local TV news segment—a report that would lead to trouble with his police department back home.

After two more days of searching, Symington returned to Nova Scotia only to be fired ten days later. The department said Symington's trip to New York was unauthorized, and that it contradicted his claim that, because of a work-related stress disorder, he couldn't perform his job duties.

While major press outlets weren't picking up the story of Trakr's role in Guzman's rescue, the Halifax *Chronicle Herald*, two weeks after Hall and Symington left Ground Zero, published a story, based on an interview with Joe Hall, stating that Trakr was believed to be the only dog to locate a survivor in the debris. The story, which made no mention of the second dog, or the hours that elapsed between Trakr's alert and the actual rescue—appeared in the paper on the eve of Symington's firing.

After his firing, his account of the rescue would garner more coverage from Canadian media and from several websites that had been documenting the work of search-and-rescue dogs at Ground Zero. Two websites would take up his cause, and report that ill will existed between Symington and his department because he had protested a new Halifax police department policy that, they reported, required search-and-rescue dogs to be euthanized upon their retirement. The department denies ever having any such policy; instead, a spokesman said, it allows handlers to purchase retiring police dogs for a dollar.

Dogsinthenews.com, a website that launched in January 2001 and focused on the work search-and-rescue dogs were doing at Ground Zero, published a picture gallery that included Symington and Trakr at the site and ran stories characterizing his firing as unjust and outrageous. Animalhelp.com published a lengthy article loaded with praise for Symington and Trakr, which linked to an online petition protesting the firing of Symington after he and his dog exhibited such courage at Ground Zero. Four years later, he and Trakr would receive the Extraordinary Service to Humanity Award, presented to him by Jane Goodall.

Despite leading to his firing, his unauthorized trip was worthwhile, Symington still insists.

"While I wish I could change the actions of my department, I would do it all again in a heartbeat," one website article quoted Symington. "I would save Trakr's life, and go to Ground Zero. There's at least one person alive today who I am sure is very thankful."

10.

The Copied Cat

■ ■ ■

This was no time for play.
This was no time for fun.
This was no time for games.
There was work to be done.

—*Dr. Seuss,* The Cat in the Hat Comes Back

COLLEGE STATION, TEXAS
February 14, 2002

In the first five years after the birth of Dolly the sheep—amid public fears; congressional debates; and the often hazy lines, when it came to cloning, between fact and fiction, acceptable and unacceptable, what served humanity and what threatened it—science, as is its nature, stealthily marched on.

By the end of 2001, scientists had successfully cloned five other species—not for the sake of pet owners, but for the sake of science. They'd produced genetic copies of everything from a mouse to the gaur—a huge, muscle-bound, endangered ox from South Asia that resembles a water buffalo. Though it lived only two days, it was still considered a "success."

Texas A&M, if not leading the way, was churning out cloned animals as quickly as anybody.

On September 5, 2001, the university's College of Veterinary Medicine announced the birth of the world's first cloned pigs—the arrival of which made Texas A&M the first academic institution in the world to have cloned three different species. On top of that, a university press release boasted, its researchers were "aggressively working" on three more—dogs, cats, and horses.

By then, A&M had cloned a docile Brahman steer named Chance, begetting a bull named Second Chance; an Angus bull named Bull 86, begetting a calf named 86 Squared; and a Boer goat, whose clone was called Second Addition but nicknamed Megan—with much of the research paid for by John Sperling, the wealthy Arizonan who, at least according to initial reports, merely wanted his dog cloned.

Dogs, though, were proving far more difficult to clone than any other animal. As Lou Hawthorne had said—and would again—cloning a human would probably be easier than cloning a dog.

With the arrival of the cloned pigs—five litters in all—A&M used the occasion to showcase its cloning portfolio to the world. For the barnyard press conference, it brought together the first litter of piglets—Porky, Daffy, Sylvester, Tweety, and T-Sipper (the runt of the group)—as well as 86 Squared, Megan the goat, and Second Chance.

The goat munched, the cattle mooed, and the piglets named after cartoon characters squealed, making it difficult for reporters to hear the scientists who took turns answering questions. Standing amid the menagerie, scientist Mark Westhusin informed reporters that there had been little progress on the quest to clone a dog—dogs being far more difficult because their eggs are opaque, and because female dogs go into estrus only every six to twelve months.

Westhusin held out more hope for cats: "I predict we'll see a cat within a year."

Six months later, in February 2002, Westhusin and other scientists at A&M were in front of the microphones again, to announce—also through research funded by Sperling—the birth of the world's first cloned cat, Carbon Copy, or CC for short, who had been born two months earlier, on December 22, 2001.

While cloning a dog was the original goal, publicity about the Missyplicity Project and Hawthorne's new company, Genetic Savings & Clone, had led to a flood of requests and inquiries from pet owners wanting their animals cloned. About 40 percent of those pertained to cats.

To Hawthorne, whose company began banking the tissue of hopeful owners' pets in 2000, it was clear that copying cats, if that technology could be conquered, could prove lucrative as well.

Stymied in their efforts to produce a cloned dog, the scientists saw cloning a cat as a stepping-stone that, while furthering their general cloning knowledge, might bring them closer to the original goal. Perhaps more significantly, it fit in with the Genetic Savings & Clone business plan. The veterinary scientists at A&M were walking a bit of a tightrope—nobly pursuing objective research, as they saw it, while at the same time doing so with money and direction from what was hoped to be a private, money-making enterprise.

"It basically was an opportunity that presented itself because we were getting a tremendous amount of calls from the general public, and a lot of them wanted us to clone their cats," A&M scientist

Westhusin says. "Cats and their reproductive physiology are a lot easier to deal with and manipulate than are dogs. So we basically just made a decision: Let's just take some of the money that we're spending on dogs and let's go buy us some cats. . . . We were already getting tissue samples from people wanting to clone their cats in the future. So we needed to learn how to clone cats and set that whole thing up. So that's how we got into it, basically public demand—plus it fit into the overall business plan Lou had for developing a company that provided pet cloning."

Not surprisingly, Westhusin went along with that plan. For one thing, the company was the conduit through which funding of the cloning project flowed. For another, he was, for a while, part of it. Company press releases named him as a founding partner in Genetic Savings & Clone, but he now says he only mulled an offer to become a partner, and didn't accept. He did, however, serve for a while as a paid consultant of the company.

The alliance between Texas A&M and Genetic Savings & Clone was the type that has become increasingly commonplace, uniting public university and corporate interests—taking a bubble of intellectual curiosity and institutional expertise and pumping into it that vital life force: money. No pipette is required, but strings may be attached. If the project goes successfully to term, the scientists gain recognition and the business sees its value increase. It's the kind of arrangement—awkward as it can sometimes be—that can give birth to major scientific breakthroughs, and major ethical questions.

It took a calico cat to bring it down.

The world's first cloned cat, as it turned out, looked little like her donor—a fact that would create serious implications for the business side of the alliance, exacerbate existing personality conflicts, and, probably more than any other factor, lead to the dismantling of the partnership.

The culprit was something known as X-inactivation, a chromosomal phenomenon that dictates the markings of the calico cat, explains why so few of them are male, and accounts for why women, unlike men, have patches of skin that contain no sweat glands.

Males, whether cat, dog, or human, have one X chromosome; females have two. X-inactivation is the process in which one of a female's X chromosomes gets turned off. At a point in the formation of a female—about the time they consist of thirty-two cells—one or the other X chromosome randomly shuts off in each cell. As those cells continue to multiply into the billions, some contain active versions of one of the X chromosomes, some contain inactive versions.

Just as a woman can, as a result of that, have patches of skin that don't sweat, calico cats have patches of fur in different colors, depending on which X chromosomes inactivated where. That's also why male calicos, with only one X chromosome to begin with, are extremely rare.

Scientists at Texas A&M, though they'd repeatedly shown they could create one animal from another, had no control over X-inactivation and therefore no way of knowing whether the clone would have similar markings to the donor cat. If physical replication was the goal, using a calico cat as the donor was the worst possible choice that could have been made. Cloning a gray cat would likely produce a gray cat. Cloning a calico was a crapshoot, and—given that the calico's chromosomal flukiness is used to teach even high school–level biology—there's virtually no way the scientists didn't know that.

The world's first cloned cat was born gray-and-white. The laboratory cat she was cloned from, Rainbow, was gray-white-and-orange.

To make matters worse, they named the clone Carbon Copy, calling more attention to the fact that she was not.

For the scientists, even with the variance in coat colors, CC, as she was commonly called, was a triumph worth trumpeting. To Hawthorne, who was building a company that promised to produce identical genetic copies of people's pets, it was a public relations catastrophe of gaur-like proportions.

Whether it was carelessness, the only choice available at the time, a product of the differing goals of Hawthorne and the scientists, or an intentional slight (the A&M researchers deny that), scientists plucked Rainbow from their laboratory cat collection to serve as one of two donors used in the experiment.

An adult male cat—not a calico—was chosen first. Scientists obtained cells from his mouth, isolated and treated them, then implanted them in the egg cells of other cats, which had their nuclei removed. Out of 188 nuclear transfers, 82 cloned embryos formed. Those were transferred into seven surrogate cats. One of them became pregnant, with a single kitten, but the fetus ceased to develop and was surgically removed. For a second attempt, scientists used Rainbow, and used cells taken from her ovaries. Through nuclear transfer, five cloned embryos resulted, which were all placed into a surrogate cat named Allie. One of them, CC, survived to birth.

CC didn't resemble her donor physically or behaviorally. In addition to the difference in coat colors, CC was sleek and playful, while Rainbow was chunky and reserved, though those distinctions could be attributed to age and environment. The differences were no big deal to the scientists—they'd made a point all along of saying they weren't re-creating the same animal, just one with the same genetic makeup. But Hawthorne was livid.

"A&M's attitude was they liked the damage that was done to Genetic Savings & Clone," Hawthorne says. "They were happy about the fact that it caused us problems, because they were never comfortable with the commercial side of the business in the first place. Their whole media strategy and response to that event was to announce, over and over and over again, 'It's reproduction and not resurrection,' when, from a genetic standpoint, it is resurrection. And the only reason it didn't look like resurrection is that they switched the genetic donor in the lab and went with a calico instead of the one that we had picked out. That was sort of the beginning of the end as far as our whole relationship."

Seven years later, Hawthorne still seems to stew about it, while the scientists now, as then, calmly dismiss his concerns and his implication that they took some delight when donor and clone came out as a mismatch.

"We don't quite agree," says Duane Kraemer, an A&M scientist who speaks so quietly as to sometimes grow inaudible. He denies that a calico cat was purposefully chosen, and says the difference in coat colors, while it may have detracted from what Hawthorne was trying to achieve, in no way blemished their success.

"She wasn't selected because she was calico, she was selected because she was the next logical donor—just the next cat available. We knew about X-inactivation, but we didn't know whether it would be reproduced or not." To the scientists, he says, that didn't matter.

"On balance, it's probably a good thing—it was probably fortunate that it turned out this way, because people had the idea they were getting their animal back, and this made it easy to demonstrate that wasn't what was happening. But it wasn't done intentionally for that reason," Kraemer said. "When you market things, you should market in full truth. . . . Lou's just learning that."

CC, who spent her first year in a laboratory cage, now lives with Kraemer, who—as the one who brought her into the world, and almost as if making up for her incarceration—went on to supply her with her own house and family.

Kraemer and his wife, who live on seven acres in Bryan, Texas, bought CC a husband, a gray cat named Smokey. They had a litter of three, the natural way—Tess, Tim, and Zip—and the entire cat family now lives in their own two-story house, twelve feet wide by twenty feet long, with heating, air-conditioning, plumbing, hardwood floors, lofts, and a screened porch, all constructed for them by Kraemer.

The other cats used in creating CC were both adopted—as the Missyplicity Project sought to do with all its dogs and cats. Rainbow, the donor, went to a home in Austin, but later died from tumors. Allie, the cat who served as surrogate mother, was adopted as well, but was later struck by a car and killed.

CC, at age eight, has enjoyed good health and shown no signs of having a shortened life span, says Kraemer, who's now working on cloning birds. He and his wife, who also have two dogs, still get requests from the curious, asking to see the world's first cloned cat. Generally, when they arrive, it's anticlimactic, Kraemer says. "They see it's just another cat."

The science world wasn't all that wowed by CC, either. "CC was not created to advance medical knowledge or provide fundamental biological insights," the journal *Nature Biotechnology* editorialized after her birth. "She was created because there is a market among certain rich cat owners for resurrected animal companions."

While Texas A&M scientists would continue trying to clone a dog for another year after CC's birth, those efforts ceased when the relationship with Genetic Savings & Clone ended in 2003. Over five years, using 225 dogs, they had produced only one pregnancy, and it didn't come to term.

Genetic Savings & Clone relocated to Wisconsin, taking a few key scientists from A&M along, to continue the effort on its own. In a business complex outside Madison, GS&C scientists used canine eggs provided by spay/neuter clinics in exchange for contributions, and surrogates procured from a network of breeders—dogs that, because of shortcomings, couldn't be sold on the market. They continued trying to achieve a successful dog cloning, and began filling orders for cat clonings.

But not calicos.

While CC was the biggest factor in bringing the partnership between A&M and GS&C to an end—a playful, four-legged reminder of the differing goals of the parties involved—other forces were at play as well: GS&C's frustration with all of Texas A&M's bureaucratic hoops, and the lingering discomfort among the veterinary school's scientists with the marketing side of the project.

"I got a little crossways with the idea of how you would market it," says Westhusin. "I talked to so many clients and people on the phone who were just distraught over their pet dying, and were thinking that cloning was a way to resurrect their pet. What I didn't like about the whole concept was that some pet owners, because they were so emotionally involved with it, could not separate the issue of reproduction versus resurrection. And I was afraid that the business side, being forced by the need to make money, would fall too much over on the side of taking advantage of the emotional needs of people and their attachment to their pets.

"I couldn't in good conscience be a good salesman for it, I guess," he adds. "I just wanted to make sure there was full disclosure and that they understood they were not getting their pet back. It may look different and it may act different. We just don't know. There's a lot of evidence to suggest that many of them do look and act like their originals. But there's also evidence to suggest that a lot of them don't."

Personality conflicts played a role as well, Westhusin admits. The lead scientists were quiet, nose-to-the-grindstone, eye-to-the-microscope types, prone to keeping mum about their research until the time came to write a paper about it. Hawthorne, on the other hand, was prone to making the most of the news media.

"Lou and I got into a lot of disagreements and fights," Westhusin says. "One of the things he didn't understand that I did—because I had to take the brunt of the crap—was that you should just do your stuff and keep your mouth shut.

"There were three major things that drove this research," Westhusin explained, "and the importance of those three things would differ depending on who you asked. One, John wanted to clone Missy, and he could afford it. Number two was the commercial potential for pet cloning. And number three was the things we could learn in terms of basic research about the reproductive physiology in dogs and applying it to models for human disease.

"If you asked me which one of those was most important, I would say the research. The funding allowed us to do research, publish papers, learn more, provide jobs, and pay graduate students—all the things that are related to my mission here as a professor. For Lou, it would be the commercial part, creating a successful company. And for John, it was, absolutely, 'I want Missy cloned.'"

But with nearly five years having elapsed since the Missyplicity Project began, with a calico cat having dealt GS&C a serious blow, and with the connection to A&M severed, Sperling—a man not used to waiting for what he wants—would have to wait a little more.

11.

The Forever Pet

■ ■ ■

When my time comes, just skin me and put me up there on Trigger, just as though nothing had ever changed.

—*Roy Rogers*

KEYSER, WEST VIRGINIA
2002

Less than loved in town and often at odds with her parents, Bernann McKinney didn't hesitate when—after her farmhouse in the mountains of North Carolina was seriously damaged by flooding—her pen pal Elliott Brown suggested she come to California and move in with him.

It took three vehicles to make the move. She hauled three horses and three dogs, and the journey took three weeks, the delays a result of her firing drivers and having to hire new ones along the way. Booger, of course, rode with her.

Tough Guy, the mastiff whose attack left her disabled and who died shortly thereafter, didn't make the trip—though his form still lingered.

After her phone call to Ian Wilmut, the scientist who cloned Dolly the sheep, McKinney had decided on another way to keep Tough Guy around, one she'd resorted to with other dogs that had died—she'd put him in the freezer, with plans to have him mounted.

When Brown came East for a visit about six months after the 1996 attack—the first time they'd met in person—McKinney persuaded him to help her take Tough Guy for an autopsy, then to a taxidermist, ensuring the dog who mauled her would, in a way, be around for eternity.

McKinney, who was partially mobile by then, joined Brown for a multistate trip (with Tough Guy's corpse in the back of the station wagon, packed in ice) in search of a veterinarian willing to do an autopsy and a taxidermist willing to stuff the dog. When the mount of Tough Guy came back, it was misshapen and didn't really resemble what he looked like in life. Still, she'd keep the mount, at least until her move to California. With so many other belongings and live animals to transport, Tough Guy didn't make the cut and would be left behind in her empty farmhouse.

Long before dog cloning became possible, as far back as ancient Egypt, humans have sought to memorialize their departed dogs—from mummifying them to posting modern-day Internet tributes, from preserving their memory to preserving their corpses, most often via stuffing and newer renditions on that theme.

Around the time Texas A&M was cloning the world's first feline, Chris Calagan, a Bible college teacher in West Virginia, and his wife, Sandy, a bookkeeper, started looking for ways to hang on to their seventeen-year-old calico cat Naomi, or at least their memories of her. The Calagans paid a visit to a taxidermist. They weren't interested in traditional taxidermy, in which an animal is skinned and its pelt stretched over a mount. Skinning Naomi, even once she died, was just too gruesome a thought. And that was just as well, because traditional taxidermists have never been too keen on pet owners as customers.

For one thing, the typical taxidermist's skill set doesn't usually include hand-holding and grief counseling. Usually, their customers are bringing in a creature they have slain, not one they coddled and treasured. For another, their shops lacked pet-size molds, though the bobcat form would sometimes come close enough to suffice. On top of that, any pet-owning customer was likely to be disappointed. Capturing the essence of a pet—an animal a family had shared a home with for one or two decades, as opposed to one briefly glimpsed in the rifle sights—was difficult, if not impossible. Pet owners, seeking a remembrance of a creature full of life and overflowing with spirit, often weren't satisfied with the motionless, glass-eyed statue that was returned.

Looking for something more tangible than photographs and less creepy than a pelt mounted over plastic, the Calagans stumbled upon some 1970s-era technology that, by 2000, had spread far beyond the Smithsonian, where its use on animals originated, and was turning up in the workshops of some of the more high-tech practitioners of taxidermy: freeze-drying.

Rather than purchase the service, they purchased the machine and, in 2002, set up shop in a shed next to their house in Keyser, West Virginia, about twenty-five miles due west of Paw Paw. Perpetual Pet was open for business.

Not to be confused with ForeverPet or PerPETuate—two companies that, before cloning was achieved, were offering gene banking for pets—Perpetual Pet returned to customers the same animal, albeit one lacking any internal organs or, for that matter, life.

The Calagans kicked off their pet preservation business by freeze-drying Naomi, who had hung on to life until after turning eighteen. They were pleased with the results. "It's really her, and she looks just like she always did!" the Perpetual Pet website says. "How wonderful knowing she will always be with us. We are thrilled that we can help others who are grieving the loss of their pets, giving them the opportunity to see and touch their pets again."

Since then, Perpetual Pet—now among dozens of companies, mostly taxidermists, offering the service—has freeze-dried more than a hundred pets for pet owners, many of whose testimonials appear on the website.

"We were so distraught when our darling Jenny passed away," wrote a Wisconsin couple. "We did not want to bury her and lose her forever, and we did not want to have her 'outsides' put on over a form as they do in taxidermy. We wanted our whole kitty to be with us forever." Jenny's body is now curled up in permanent "near-sleep" position, eyes almost shut, snug in her bed. "We love having our dearest little one with us in our home. Seeing her each and every day, sleeping so sweetly, has eased our anguish over a life that was much too short."

The testimonials go on—from the owners of Sinbad, a sweater-wearing Chihuahua in Canton, Ohio; Zak, stretched out in his dog bed in Taylor, Texas; Cisco in San Diego, California, a freeze-dried dog who sits atop a table, wearing headphones and sunglasses; and Trigger in New York City, a black cat freeze-dried in the sitting position, an outstretched paw reaching for a spool of thread.

Freeze-drying is just one of many ways pet owners seek to preserve an animal's legacy. For those with purebreds, it has always been as easy as breeding their favorite dog to get a repeat version. Those whose pets are among the masses of spayed and neutered mutts have resorted to other routes to keep their dog's memory alive, ranging from mild to extreme. In modern times, it's not viewed as too quirky to hold on to a favorite pet's cremated remains, a lock of hair, a favorite toy, or the bowl inscribed with his name. It's fairly common, upon a beloved dog's death, to go out and get another that looks just like him.

Nor is it deemed so wacky anymore to hold a formal funeral. Just as some pet owners will do all in their power to extend a dog's life, there are those who will do all in their power to preserve that dog's memory; and there are more options for that than ever for the bereaved pet owner.

Pet funerals, for which the backyard once sufficed, can now be custom-ordered, complete with viewings (after a postmortem grooming), engraved headstones, formal services, and burial or cremation. Once commonly performed en masse, individualized cremation can now be requested, with the ashes returned inside a variety of forms—a porcelain urn that resembles the original dog, a hollowed-out rock that can be laid in his favorite spot, or a "digital photo urn." It's a boxy frame, available in oak or walnut finish, which contains one compartment to hold the ashes of a dog up to 75 pounds, another to hold the memory stick on which you can load 256 gigabytes worth of photos of your pet. (Remote control included.)

There are companies that will make diamond jewelry out of the carbon in your pet's ashes, and others that produce jewelry designed to hold those ashes, such as a pendant one can wear around one's neck. Or you can place your dog's ashes inside a stuffed animal especially designed for that purpose, allowing you to cuddle your pet's cremains long after its death.

"You need something to hold," explains Alexandra Lachini, a California psychic who created a company called Huggable Urns. "It's like when your boyfriend breaks up with you and you go and

The author meets Snuppy, the world's first dog clone, at Seoul National University's veterinary school. *(Rona Kim)*

James Symington and Trakr at Ground Zero, where, Symington says, his dog found the last surviving victim of 9/11, Genelle Guzman. *(Associated Press)*

Symington meets the five clones of Trakr—Trustt, Solace, Valor, Prodigy, and Déjà Vu—at a Los Angeles press conference. *(John Woestendiek)*

Lou Hawthorne, CEO of BioArts International, and his dog, Mira, who was cloned from his mother's dog, Missy. *(John Woestendiek)*

Mira at the dog park. Like the husky–border collie mix she was cloned from, Mira has one erect ear, one floppy ear. *(John Woestendiek)*

The exterior of RNL Bio's offices in Seoul. The company was the first, not counting Hawthorne's contracted cloning of Missy, to deliver a canine clone to a customer. *(John Woestendiek)*

The original Chance was part of Ralph Fisher's traveling show, in which customers could be photographed while sitting atop the Brahman bull. Chance also appeared in a movie, at presidential inaugurations, and on David Letterman's show. *(Courtesy Ralph Fisher)*

Above: Fisher hugs Second Chance. The bull clone, who displayed many of the original Chance's habits, lived less than nine years. *(Courtesy Ralph Fisher)*

Right: Ralph Fisher thought the clone of his bull would be as sweet and gentle as the original. Here he displays the jeans he was wearing on the two occasions that Second Chance, the world's first cloned bull, attacked him. *(John Woestendiek)*

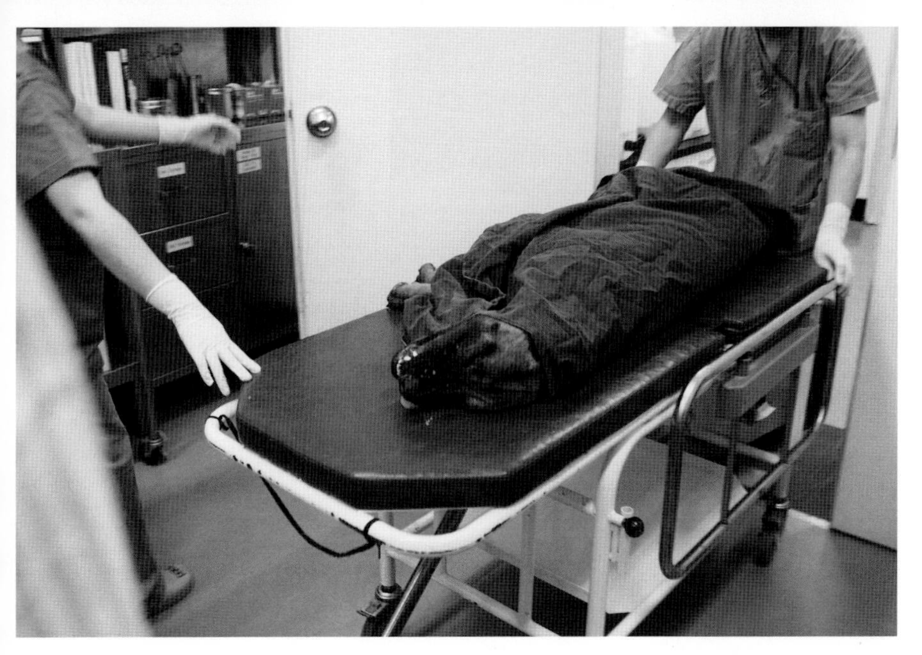

The first of two dogs to have their eggs harvested for use in cloning an auction bidder's dog is wheeled out of the operating room at Dr. Woo Suk Hwang's Sooam Biotech Research Foundation. *(John Woestendiek)*

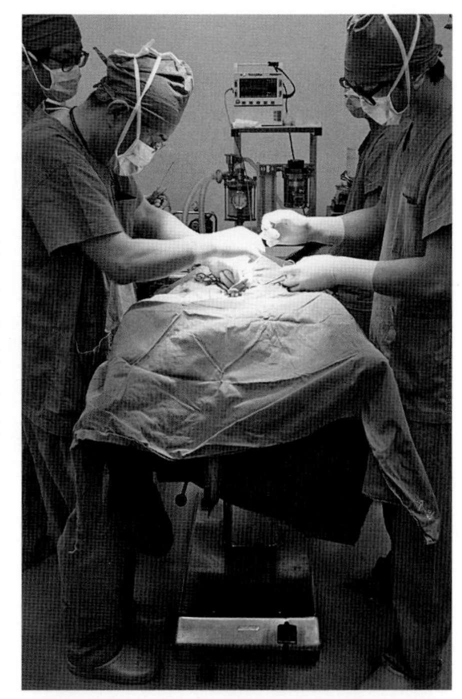

Dr. Woo Suk Hwang and his team at Sooam search for egg cells in the uterus of an anonymous donor dog. Seven were harvested, enucleated, then injected with skin cells of the donor dog, the fifth to be cloned in connection with the BioArts online auction. *(John Woestendiek)*

The world's first commercially cloned dogs were genetic copies of a stray pit bull that Joyce Bernann McKinney found wandering on the side of a road. Six months later, Booger came to McKinney's aid when she was attacked by another dog, and went on to become her service dog. *(Courtesy RNL Bio)*

The Booger clones shortly after birth at Seoul National University. *(Courtesy RNL Bio)*

Joyce Bernann McKinney holds up one of the Booger clones during the trip she made to South Korea to meet them. *(Courtesy RNL Bio)*

As of early 2010, Peter Austin Onruang, who operates a spy gear shop in Los Angeles, was awaiting the clone of his Yorkshire terrier–schnauzer mix, Wolfie, who died in 2009. (Courtesy Peter Austin Onruang)

Onruang, as an RNL Bio customer, is having Wolfie's sister, Bubbles, cloned as well. He says his hope is that the original dogs' souls will return to the bodies of the clones. *(Courtesy Peter Austin Onruang)*

Dogs waiting to be bought and butchered at Moran Market, a block-long open-air bazaar outside Seoul. Easy access to dogs, and few restrictions on their experimental use, helped South Korea corner the market on dog cloning. *(John Woestendiek)*

grab his old shirt and wear it around. Or like when babies grab their little blanket. It's a natural instinct for us." Lachini's Internet-based company makes pillows and stuffed animals with a zippered pouch, into which one can insert a plastic bag of their lost pet's ashes, or, for that matter, a human's. When Oprah Winfrey's dog Gracie choked to death, Lachini made the talk-show host a huggable urn, but was unable to get her to accept it. When Winfrey lost a second dog, Sophie, to kidney failure in 2008, she tried again. When John Travolta's son died, she sought to provide him with one as well.

The Internet provides a host of other opportunities to grieve and memorialize a pet, including websites where one can stroll through rows and rows of virtual tombstones and epitaphs, reading owner tributes. Virtualpetcemetery.org is just one of dozens of websites to which bereft pet owners turn, partly for the therapeutic effect of recording their remembrances, partly to let the rest of the world know how special their pet was. Most of the eulogies left by pet owners, at $20 a shot, are intimate and heartfelt accounts, at once anthropomorphic and anguished:

"Shiloh (My Handsome Prince) and I adopted each other in July of 1997; he was 18 months old, I was 42 years old. We shared our lives for 11½ years. He passed away September 25, 2008, from complications associated with lupus. Both of us came from incredibly abusive relationships and welcomed the chance to have a new life with someone whom we could trust. . . . Shiloh was my companion. I was a better person when I was with him. I was kinder when he was with me. . . . He gave me reason to live when I had none. He loved me when no one else did. He believed in me when no other would. . . . My only consolation is that I know that he will be the first person to greet me in Heaven—as long as I live worthily. That is my eternal goal, to return to Heaven and be with my precious boy."

Many of the tributes are written directly to the pet, as if there is Internet access in doggie heaven, and dogs are able to log on.

"Dear Jumbo, Our beautiful Son . . . If you could only know how much joy you brought to our lives . . . Please forgive me for the wrong I have done. The nights I didn't come home. The drugs and alcohol. You made my life worth living. My heart hurts so deeply, and I feel empty even as the days go on. Poppy misses you as well. . . . Grandma told me once we received you back in the urn that I would feel better. I don't feel better."

The accounts go on and on, in almost relentless despair. Most of the pet bereavement websites, while directing visitors to resources and providing them with a chance to put their feelings into words, simultaneously seem to reinforce the all-enveloping shroud of sadness.

Visit petloss.com and you'll hear the sullen strains of the song "Evergreen" in the background as you examine options that include chat rooms, support groups, message boards, recommended reading, and laminated copies of inspirational poems for a $50 donation. Lightning-strike.com's original webpage opens with a flash of lightning, a reminder of how out of the blue a pet's death can be, which leads into a version of the Beatles' "Yesterday." It offers "a cybershoulder for grieving pet owners"—books, website links, poems, and more, all interspersed with ads for gravestones and caskets and cremation jewelry.

And for those who find comfort in their pet's continued physical presence—a likeness of the entire animal—there are companies like Perpetual Pet.

It takes about two months for the Calagans to freeze-dry a Chihuahua, about three months for a cocker spaniel. A medium-size dog—forty pounds is their limit—takes four months. Most of the animals arrive at the Calagans' home in Keyser in a frozen state, shipped overnight by FedEx. Up until then, ideally, the corpse has been kept on ice—either at a veterinarian's office, or, for those whose pets died at home, in the family freezer, which Perpetual Pet recommends.

"When we're ready to work, we thaw the pet out," Calagan says. The pets' internal organs are removed and it is washed and then posed according to the owner's instructions. The pet is then refrozen and put in a vacuum chamber, which gradually draws all moisture out of the body. Over the next few months, Calagan or his wife will check on the animal's status to make sure it is holding its pose, or is in need of minor corrections.

While some will continue to stroke and talk to their freeze-dried pet, Calagan doesn't think his clients are mired in denial. "Everyone knows it's not 'alive' anymore," he says. "But having its body there, having its appearance essentially unaltered, to some extent, it's like part of the pet is still with them. It's hard and stiff to the touch, and much lighter, and doesn't move, but people can put them on the spot where they used to sit and see them and touch them and pet them. Just being able to retain their pet, at least the part of it that's physical, that's the largest thing."

Using taxidermy to preserve pets rose in popularity in nineteenth-century England and France, as did pet ownership. Keeping a likeness of a pet that had passed on became, in some circles, a rising fashion. For many, a portrait served as memory enough, and often, in Victorian times, those would be painted after a dog's death. In posthumous pet portraiture—first in paintings, later in photographs—animals would be arranged to look as if they were curled up for a long nap. Eliciting many more snickers—the same sort of snickers one hears today when it comes to cloning—was the choice of stuffing one's beloved dog, cat, or bird. Like cloning, it was generally viewed as the act of an eccentric with more money than he knew what to do with, whose love or esteem for his animal exceeded "normal" bounds.

"This mode of remembrance repulses me," French author Alfred Bonnardot wrote at the time. "It is a sad thing to see one's little companion whose look was once so lively and bright forever immobile and staring. Moreover, if one kept all his successors this way, one would end up by having a somewhat cluttered and encumbered museum."

As with the science of cloning, the craft of taxidermy had its own "rogues," outcasts to whom making a "realistic" mount came second to making a statement, preferably humorous. In the 1930s and 1940s, fantasy taxidermy took hold, resulting not just in re-created animals performing human acts—poker-playing rabbits, chipmunks in tuxedos—but "chimeras" as well, commonly made by grafting the head of one animal onto the body of another. Perhaps the most famous of these is the "jackalope"—a rabbit with antelope antlers. Who fashioned the first jackalope, apparently intended to pull a fast one on any city slickers passing through town, is a mystery, but plenty of jackalopes remain.

When cowboy star Roy Rogers's horse, Trigger, died in 1965 at age thirty-three, the Rogers family had him mounted, his skin stretched over a plastic mold, posed proudly in the position of a horse at its liveliest—reared up on its hind legs. Trigger became the main draw at the Roy Rogers–Dale Evans Museum. The family also had Dale Evans's horse, Buttermilk, and their German shepherd, Bullet, mounted to become museum pieces. Rogers, before his death in 1998, joked about having his own body stuffed and placed atop his rearing horse, but he never actually pursued that.

Still other humans have sought to preserve their pets, and

themselves, through mummification—not so they can be returned to life as they knew it, but to ease their passage into the next world.

In Salt Lake City, members of Summum, another religion that credits its existence to space aliens, are mixing ancient Egyptian rituals with high-tech techniques to preserve the bodies of deceased pets. Since the late 1980s, they've mummified more than a hundred animals, including Maggie, a standard poodle that belonged, and still does, to Su Menu, a fifty-eight-year-old piano teacher in Salt Lake City.

"I just didn't feel good about cremating her and putting her in an urn, or just putting her in the ground," Menu says. "I did it partly to memorialize her life, because she had been such a good dog and companion to me. . . . I want to be mummified, so I wanted to do the same for her. I want her with me in my next life too. . . . We'll continue on and her soul will be with mine, in whatever forms we're in."

In 1993, after Maggie died at fifteen, Menu, a church member, turned the dog's body over to Ron Temu, chief "thanatogeneticist" for the church. Temu and his assistants bathed the body, removed its blood and internal organs, and immersed it in a vat filled with a chemical preserving solution, where it remained for about ten weeks—about the same amount of time Egyptians took to dry a body with salt. After that, Maggie was bathed again and covered with lotion. Then numerous layers of cotton gauze were wrapped around the body and a polyurethane membrane was applied over the gauze, forming a permanent seal. That was followed by a layer of fiberglass and resin.

The process takes six to eight months, and involves about a thousand hours of labor. In the final part of the process, the wrapped and sealed pet is encased in a mummiform, usually made of bronze or stainless steel, which is also sealed with resin. For Maggie, Menu chose bronze. It was two years in the making, crafted, like all Summum mummiforms, by Stan Watts, an artist most famous for a forty-foot-high, 17,000-pound bronze monument that portrays three firefighters raising the American flag at Ground Zero.

The Summum's practice—what the group's members call the "Rites of Transference"—is based on their belief, in line with those of ancient Egypt, that a body must be preserved to allow its soul to travel to its next incarnation. Those beliefs, their tax-exempt mummification factory, and the fact that they also operate a vineyard, raise some eyebrows in predominantly Mormon Salt Lake City.

About fifteen hundred humans have signed up for what the church calls the "Permanent Body Preservation System," at a cost of $67,000 (not including mummiform)—among them football players looking to be mummified in athletic poses, military men seeking to be mummified in uniform, and a radio-talk-show host who wanted to be holding a microphone for eternity. But Temu, who started work on his first human mummy in 2009, has kept busy with pets.

In addition to easing a pet's passage into the next world, he says, both the process and the final result serve as solace to pet owners who remain, for the time being, in this one.

"Quite a few of our clients get back to us and tell us what they've done with them, like putting the cat where it always used to sit, at the top of the stairs, where it can see what is going on," he said. "A lot of them don't tell friends that there's an animal inside. They make up a story like, 'I saw this, and it reminded me of Max, so I bought it, and I call it Max, and I put it where Max used to lie.'"

12.

Good-bye, Dolly

■ ■ ■

> And of every living thing of all flesh, two of every sort shalt thou bring into the ark . . . they shall be male and female.
>
> —*Genesis 6:19*

EDINBURGH, SCOTLAND
February 14, 2003

In the second week of February, Ian Wilmut, the leader of the team that created Dolly the sheep, sat in his office at the Roslin Institute, holding his head in his hands.

He had just gotten word from the Scottish College of Agriculture that the recent hacking cough Dolly had developed was, as feared, pulmonary adenomatosis, a common disease in adult sheep, usually signifying the beginning of the end. A CT scan had confirmed her lungs were full of tumors.

As with the arthritis Dolly had developed at age five in her left knee, there was no way to know if the tumors were connected to her being a clone. But the lifestyle she had lived as the first mammal cloned from an adult cell—nearly permanently enclosed and on display—may have made her more prone to the sickness, Wilmut admits in his book *After Dolly*.

"Perhaps if Dolly had been allowed to roam the green fields, gulp in fresh air, and gaze at the Pentland Hills, she would not have succumbed," he wrote. "But, because of her celebrity status, she had to be protected from lunatics who would want to harm her, criminals who wanted to kidnap her, and local students who might pull pranks."

With no chance of a cure, and with Dolly having difficulty simply recovering from the anesthetics she had received, Wilmut didn't have to think long. At three-thirty p.m. on Valentine's Day, 2003, he ordered the cloned sheep that had brought him and the institute fame and fortune euthanized with a lethal injection of barbiturates. She had not yet reached seven years of age.

The same afternoon, she was skinned by taxidermists from the Royal Museum of Scotland. The Roslin Institute had made an agreement, when Dolly was just two, that she would be preserved and exhibited at the museum upon her demise. Quickly, to avoid the possibility of decay, her skin was removed, pickled, tanned, and stretched over a fiberglass mold of her body. Glass eyes were inserted in the head, and a nose was fashioned out of plastic.

The rest of her remains went to the laboratory to be studied in an attempt to determine what role, if any, being a clone played in her early death. Most sheep live to eleven or twelve. The postmortem exam revealed that the telomeres in Dolly's cells were shorter than those that would normally be found in a sheep of her age. Telomeres shorten with age, and Dolly's were those of a much older sheep.

Despite that, Wilmut insists that Dolly—even with her arthritis and early death—was "unscathed by cloning."

Dolly's premature death did nothing to slow down animal cloning. If anything, it seemed to speed up that year. In May, the first horse was cloned, in Italy, followed by the first mule, a joint project of the University of Idaho and Utah State University. In July, Japanese scientists announced they planned to attempt to clone a woolly mammoth, using remains found in the Siberian tundra. In August, Gerald Schatten, a scientist from the University of Pittsburgh, was granted $6.4 million by the National Institutes of Health to attempt to clone a monkey. In September, the first rat was cloned, in France, and in December, the first cloned deer's birth was announced, back at Texas A&M.

In the labs of Genetic Savings & Clone, work continued on cloning the world's first dog—despite the Roslin Institute's warning, in a 2002 report, that "cloning will not 're-create' a loved pet. A clone might be 99.95% genetically identical to the original but it will grow up with a personality and behavior all of its own."

Even with such precautionary notes from Wilmut and other scientists, cloning—thanks in large part to Hollywood, science fiction, the news media, and sheer repetition—was still being viewed by many as reincarnation. When humans hear a story often enough, whether it's labeled fact, fiction, or something in between, they tend to believe it.

Take Noah's Ark. Once built, it was to be loaded with two of every species of animal, one male, one female. That way, God explained, once He purged the earth of corruption, once the floods washed away the "wickedness of man" and every "creeping thing that creepeth" upon it, Noah and his floating menagerie could, when the waters subsided, go forth, be fruitful, and multiply. As the book of Genesis makes redundantly clear, that's the way God planned it, and that was the conventional wisdom for the next seven thousand or so years: it takes two animals, male and female, to reproduce another one.

Just as the ark legend has failed to hold water under scientific scrutiny, the God-given tenet that a mother and father are needed to produce offspring was turned on its head in the summer of 1997, when Wilmut and his team, working at a quiet laboratory nestled in the deep green countryside seven miles outside Edinburgh, Scotland, announced the birth of Dolly the sheep.

Dolly wasn't the first animal cloned, or even the first sheep cloned, but she was the first mammal cloned from another adult mammal—the first proof that all God's creeping creatures, likely even man, could be replicated not just from an embryo cell but from adult cells as well. As Baby Louise, the first child conceived in a test tube (actually a dish), had shown twenty years earlier, the sex act—or even the parents' presence—was not required for conception. Dolly took it a step further, showing that, when it came to being fruitful, the male of the species wasn't mandatory either.

That animals could be cloned had been proven more than a hundred years earlier—first by Hans Driesch, a German biologist and philosopher who, while working for the Marine Biological Station in Naples, performed the first human-assisted animal cloning as easily as one makes salad dressing.

In 1894, Driesch put a two-celled sea urchin embryo in a beaker of sea water, shook vigorously, and watched in amazement as each shaken-apart section became a new sea urchin. He repeated the process with a four-celled embryo, and ended up with four complete sea urchins. Low-tech as it was, it was man's first cloning of an animal, and it disproved the prevailing theories of the time—most of them put forth by August Weissman, whom Driesch had studied under—that cells had distinct, limited, preassigned missions.

Weismann, whose interest in the natural sciences stemmed from collecting butterflies when he was a teenager, believed that traits acquired in life are not passed on to offspring. That ran contrary to an increasing number of anecdotal claims at the time, such as reports of cats who had lost their tails in accidents birthing litters of tailless offspring, and of Jews being born without foreskins. Weismann also believed that once a cell started dividing, each new cell had a predetermined mission, and that as a cell continued to differentiate, there was a steady loss of hereditary information, or what would later become known as genes. To prove his first point, Weismann whacked the tails off rats, observing the offspring the tailless parents would go on to produce. He did this for twenty-two rat generations—lopping off 1,592 tails in all—and found no rats born without tails.

Driesch, with his sea urchins, would reroute conventional thinking, and disprove Weissman's theory that all cells had specific, unalterable missions. The sea urchins seemed to show that cells, despite the specific jobs they go on to do after dividing and dispersing, don't lose their "totipotency," or their ability to serve as jacks-of-all-trades, forming any and all kinds of tissue.

Another German would take cloning—though it wasn't being called that yet—to the next level. Hans Spemann, the son of a bookseller, after studying biology at the University of Heidelberg, came down with tuberculosis. One of the books he took with him to read while quarantined in a sanitarium in the winter of 1896 was Weismann's *The Germ-Plasm: A Theory of Heredity*. Spemann also had access to a little-known paper by Gregor Mendel—one that had gone unnoticed for decades.

Mendel was the abbot of a small monastery in Austria, where he pursued his interest in the inheritance of physical characteristics, mostly via bees and peas. In the monastery garden he tended, he noticed some peas flowered white, some purple. He began a series of cross-breeding experiments, keeping meticulous records of all his results. In the first round, he was surprised to find the purple and white flowering pea plants, when crossed, resulted not in the lighter shade of purple that he expected, but were instead all purple.

Working with subsequent generations, he would identify recessive traits and dominant traits, and conclude that particles inside cells—"*Elemente,*" he called them—shaped what peas, and humans, would become, and that both parents contributed to the mix. Even though Aristotle in the third century B.C. had suggested both parents contribute life-forming material during reproduction, popular thought into the nineteenth century was that sperm was the determining life force, and women merely served as incubators. Mendel's paper "Experiments on Plant Hybrids" was published in an obscure journal in 1865 and received little attention until the turn of the century. It wasn't until after his death that Mendel's work would become the foundation of modern genetics.

Spemann, his tuberculosis behind him, delved deeper into the microscopic world of cells while at the University of Würzburg—using the knowledge provided by Mendel, Weismann, and others, hundreds of eggs provided by salamanders, and some special tools of his own creation, including a hair from his baby daughter, tied into a noose.

A master craftsman, Spemann built his own tools. He would hold glass rods over a burner and stretch the glass into delicate, wispy strands, solid enough to maneuver cells under the microscope. But even those didn't work well on salamander eggs, whose slipperiness left him frustrated. It was while holding his baby daughter, Margrette, that Spemann had an idea. Rather than trying to pick a slippery cell apart with his fragile wisps of glass, he would use a hair to tie the cell off into separate sections. An adult hair would have been too thick and rough. So Spemann snipped off a lock of his nine-month-old daughter's hair and returned to the lab. Hunched over the microscope, he used tweezers to tie one of Margrette's blond baby hairs into a noose, slipped it over a newly fertilized salamander egg, and began tightening it until the two halves, though still in the same membrane, were separated.

Then he waited. Under Weismann's reasoning, each side would grow into something different—likely a lone body part or a half-salamander. But in just a few days Spemann saw he had created two independent, fully formed salamanders. Spemann called the process "twinning," and he would fine-tune it for the next twenty-five years. He produced a few salamonsters along the way—rather than twins, a single salamander with what looked like a second belly—which he kept in the name of research.

In 1928, four years before Aldous Huxley's *Brave New World* was published, Spemann tried an even more delicate experiment. His own children grown, Spemann turned to his co-researcher, Hilde Mangold, who gladly provided him with hair from her newborn son, Christian. This time he looped the hair noose around a salamander egg in such a way as to keep the nucleus on one side, leaving the other side of the dumbbell-shaped blob empty except for cytoplasm. Once the nucleated side divided four times, Spemann loosened the noose, allowing one of the sixteen nuclei to slip over into the empty, or "enucleated," side. Cell division began on that side as well, and continued after he tightened the hair noose enough to sever the two embryos, both of which became salamanders.

It's considered by some the first cloning by nuclear transfer, though it was more like moving to the other side of the room than an actual relocation, and it was the forerunner of the technique that eighty years later—much refined, and requiring no baby hair—would be used to clone dogs.

In 1937, two years after receiving a Nobel Prize for his work, Spemann left the university and published a book, *Embryonic Development and Induction,* which both outlined his research and proposed what he called a "fantastical experiment"—using an adult cell's nucleus, transferring it into an egg cell, and producing a clone. He even foresaw the possibility of the technology being used to create a human, or at least human tissue, and being used to fight disease. That led to an uproar in the Catholic Church, which saw his research as revolting, sacrilegious, and scary. Spemann was still being compared, by some, to Dr. Frankenstein when he died of heart failure in 1941.

Two Philadelphia scientists picked up where Spemann left off. In the early 1950s, they undertook the cloning of a northern leopard frog at the Institute of Cancer Research, now known as the Fox Chase Cancer Center, in hopes of furthering their knowledge of cell differentiation. Robert Briggs's original request for funding was rejected by the National Cancer Institute as a "hare-brained scheme with little chance of success," but he persisted, received his grant, and, with fellow researcher Thomas King, began an effort to transfer the nuclei of somatic cells into those of enucleated frog eggs.

Using a pipette, Briggs and King sucked nuclei from the donor cells and inserted them into the enucleated egg cells. Of the 197 transfers, 104 began developing, 35 became embryos, and 27 grew into tadpoles. They were first nuclei to develop in a whole new cell, as opposed to the one they started out in. John Gurdon at Oxford took their work a step further in the late 1950s, when he gathered cells from the intestinal lining of African clawed frogs—cells that had already substantially differentiated—and transferred them into egg cells. Of 120 transplanted nuclei, seven completed the journey to become full-fledged frogs, a success rate of about 1.5 percent.

It wasn't until the 1960s that a name for what the scientists were doing was coined. The word "clone" was first used by British geneticist J. B. S. Haldane, whose first predictions of the technological future were made in his essay "Daedalus, or Science and the Future" in 1923. In 1963, he gave another ambitiously forward-looking speech, "Biological Possibilities for the Human Species of the Next Ten Thousand Years." In that talk, he first used the word "clone," derived from *clon*, the Greek word for "twig," and he warned as well of the hazards to which the process might lead. He died the next year.

Haldane's earlier projections—including the coming of ectogenesis, or developing fetuses in artificial wombs, had been used by his good friend Aldous Huxley as the basis for his science fiction classic, *Brave New World*, published in 1932. As it turned out, the first sign that we were getting anywhere close to Huxley's horrifying dystopia—the first reported cloning of a mammal by nuclear transfer—turned out to straddle the line between fact and fiction, and never found a definite home in either world.

In 1981, Karl Illmensee, at the University of Geneva, and Peter Hoppe, from the Jackson Laboratory in Bar Harbor, Maine, published a paper in *Cell*, a prestigious biology journal, claiming they had cloned, after 363 attempts, three mice. That sent researchers at other institutions off to clone their own mice, but none, initially, succeeded. No one could replicate the replication, and Illmensee offered no help to those trying to mimic his experiment. That only led to more doubts. Two years after his paper was published, members of Ilmensee's lab challenged his results. Multiple investigations were launched. While Ilmensee's reputation had been a stellar one, numerous errors were found in a University of Geneva investigation. Researchers stopped short of disproving his experiments, but labeled them "sloppy," "error-laden," and "scientifically worthless."

Among those scientists trying to mimic his experiment were James McGrath and Davor Solter, at the Wistar Institute in Philadelphia, who, despite going to arduous lengths, failed to clone a mouse. Normally, a failed experiment doesn't rate a journal paper, but because of the attention mouse cloning had received, and the questions left unanswered, they managed to publish two. Their paper in the journal *Science* proclaimed—a little prematurely, as it turned out—that "the cloning of mammals by simple nuclear transfer is biologically impossible."

After that verdict, government grant money to pursue cloning research dried up. Cloning became relegated, quite literally, to the barnyard, as the only ones interested in pursuing it were those who envisaged the purposes it could serve, and the profits it could lead to, when applied to livestock. Cloning became the domain of veterinarians whose research was funded mostly by large agricultural concerns—people like Steen Willadsen of Denmark.

Willadsen's original work at British Agricultural Research Council's reproductive physiology unit, where he arrived in 1973, involved working with frozen embryos and was funded by the Milk Marketing Board. From there, he moved on to splitting embryos, producing, à la Spemann, twins of sheep, cows, goats, and horses. Brash and easily bored, he pursued his own experiments, including some interspecies forays. By mixing cells from embryos of different species, he created the first sheep-goat, which made the cover of *Nature* in 1984. That year, he also cloned what was—assuming Ilmensee's mice weren't valid—the world's first mammal, a sheep. It wasn't announced until 1986, when he published a paper in *Nature*. Willadsen, even before his paper in *Nature* came out, left the lab to work for the Granada Corporation, one of three companies that, by the late 1980s, were racing to clone farm animals.

Cattle-cloning research was being funded by Granada, Alta Genetics, a Canadian cattle-breeding company in Calgary, and W. R. Grace & Company, which owned American Breeders Service in DeForest, Wisconsin, the nation's largest dairy cattle-breeding company. The University of Wisconsin, with its research funded by W. R. Grace, would win that race when, in 1986, Randall Prather, a Ph.D. student working under Neal First, a reproductive physiologist at the University of Wisconsin, produced what's considered the world's first cloned cow. Willadsen disputes that, saying he cloned a cow a year earlier but didn't get around to publishing a paper on it.

What drove the research, even more than rampant scientific curiosity, were the enormous prices cattle embryos could fetch—American farmers were paying $1,500 per embryo at the time. High death rates and abnormal development, such as one cow that, while still in its mother's uterus, grew to 150 pounds, twice the normal weight of a newborn, did little to slow down the quest.

On the other side of the Atlantic, meanwhile, Ian Wilmut was working with sheep—a far more affordable animal on which to experiment. Sheep were cheap—costing only 1/500 of the price of a cow. Wilmut arrived at the Animal Breeding Research Organization in 1972. The son of two teachers, he, through a gene mutation of his own, couldn't distinguish red from green. As a result, he had given up on his childhood plan to join the navy and gone on to study agriculture at the University of Nottingham, where he became interested in embryology. After graduating, he got his doctorate at Darwin College and was hired full-time at the research station.

After producing the world's first calf from a frozen embryo—Frostie, born in 1973—Wilmut's focus turned to genetic engineering, which, he has noted, was going on long before it got that name. "Genetic modification is, in one sense, as old as domestication," he would write later. "All varieties of domestic dogs, from Dalmatians to chow chows and Rottweilers, have been created through selective breeding."

In the 1980s, though, the laboratory version premiered, turning old-fashioned animal husbandry into something more akin to animal wizardry. The process—injecting genes into fertilized and developing embryos—was a highly inefficient one, especially for Wilmut, who has a tremor. Typically at the institute—renamed Roslin in 1993—of ten thousand embryos injected with foreign DNA, only three would survive.

In search of increased efficiency, and cognizant of the potential profits of "pharming"—creating animals who, injected with human genes, go on to serve as four-legged drug factories—Wilmut and fellow researchers began looking for a better way to engineer genes. That better way, they suspected, was through cloning.

While cloning had gone out of vogue since Solter and McGrath's failed mouse experiments, transgenic animals were becoming all the rage. Companies were racing to create animals that, by some estimates, would be able to produce $200 to $300 million each in pharmaceuticals—in a single lifetime.

Wilmut, whose father went blind from diabetes in his forties and would later die of the disease, began work on that project in 1982. By 1990, he had succeeded, creating a transgenic sheep named Tracy—not a clone but a transgenic animal whose milk contained an enzyme used in the treatment of emphysema and cystic fibrosis.

He was still working on cloning as well—seeing it as a way to more easily implant genes in an animal. Wilmut had produced his first cloned sheep in 1989, through the nuclear transfer of embryonic cells. But his goal remained to clone transgenic sheep.

In 1991, Wilmut was joined in his research by Keith Campbell, the long-haired son of a seed-seller who, as a boy, would bring so many frogs home that his mother had to sweep them out of the kitchen. Campbell was a strong believer in the versatility of cells and in the notion that their differentiation is reversible—that is, even a cell that has begun to divide has the ability to carry out different functions.

In 1995, they produced five lambs cloned from partially differentiated embryo cells, two of which died at birth, one of which died ten days later. The two who lived were named Megan and Morag. By agreement, Campbell was listed as first author on the published paper announcing that achievement; Wilmut would be listed first on the next one.

The next one was Dolly—the lamb that would shake the earth.

Using cells scraped from the udder of a nameless six-year-old ewe, then stored and "starved" in a test tube, the scientists inserted them into the egg cells of other sheep, 277 in all, then applied electric jolts to fuse them. At the six-day stage, they implanted the reconstructed embryos into other ewes. One took.

Implanted the last week of 1995, the lamb who would become Dolly had become a detectable fetus by mid-March, and on July 5, 1996, she was born, weighing 14.5 pounds—the first mammal cloned from an adult cell, and fantastical proof positive that grown-up cells, with proper timing and handling, can be turned into a whole new animal.

For six months after her birth, Dolly—named, since she came from a mammary cell, after buxom country-and-western singer

Dolly Parton—was kept a secret, allowing time for the all-important science journal paper to be written and scrutinized before publishing.

On February 22, 1997, the cloned sheep was revealed to the world, focusing public attention on the previously shadowy brave new world of biotech, and opening a door to earthly possibilities that were spine-tingling in both their promise and their threat. "A new era of biological control" is how Wilmut would later describe what the fuzzy-faced lamb represented. "In the twenty-first century and beyond, human ambition will be bound only by biology and society's sense of right and wrong."

It's that last part that the news media most quickly seized on after Dolly's birth was announced: "Researchers Astounded . . . Fiction Becomes True and Dreaded Possibilities Are Raised," read the *New York Times* headline. "Dolly Opens Door for Life After Death," another newspaper proclaimed. Wilmut was disappointed by the "sensationalism." His achievement should have been celebrated, not feared, he thought. He suspected at least some of the ominous pronouncements and aspersions that followed Dolly's birth were the result of American jealousy over not having accomplished the feat first.

President Clinton voiced qualms and called for an expert review. The Vatican did the same. The European Union called for an investigation and the British government considered cutting his facility's funding as Dolly led to a new wave of human-cloning fears—a wake-up call in the form of a species most often associated with falling asleep.

Wilmut did his best to allay those concerns: "People say that cloning means that if a child dies, you can get that child back," he told the *New York Times*. "It's heart-wrenching. You could never get that child back. It would be something different. You need to understand the biology. People are not genes. They are so much more than that."

Dolly, meanwhile, became an instant celebrity, feeding on her fame. She quickly learned that visitors, especially photographers, meant she would be rewarded with treats for posing. She'd trot to the front of the stall when she heard them coming and jump up on the railing—odd behavior for a sheep, Wilmut noted. At feeding time, she'd hog the food, eating her fill and then upending the trough, to the dismay of her barn mates, Megan and Morag. Being raised on bribes and displayed for visitors affected both her personality and her weight, but even though restricted to the barn, she lived a fat and catered-to life, unaware that she'd become an icon for both the "ungodly" peril and lifesaving promise of cloning.

Wilmut said that promise—using stem cells from cloned embryos—included possible treatments for heart disease, spinal cord injury, liver damage, diabetes, Parkinson's, and Alzheimer's. "By creating a cloned embryo of a patient, we can obtain a source of the patient's own cells—stem cells—that can be used to understand the disease, test treatments, and not only repair a body but regenerate it, too. They can metamorphose into any cell type." Even replacement body parts were a possibility, he wrote in the book *After Dolly: The Promise and Perils of Cloning.*

As for cloning entire humans, Wilmut remained staunchly opposed. And—despite getting a number of calls from pet owners after Dolly's birth—he saw little point, or future, in the cloning of pets: "The prospects of making exact copies of Patch the cat and Spot the dog look bleak," he wrote.

13.

Korea Joins the Race

■ ■ ■

> A dog is not considered a good dog because he is a good barker. A man is not considered a good man because he is a good talker.
>
> —*The Buddha*

SEOUL, SOUTH KOREA
2003

In the thirteenth century, during the Goryeo Dynasty of Korea—long before the country divided into two—Yun Ui Choe had reproduction on his mind.

It was a far-fetched dream—to take one character and make possible the infinite duplication thereof. Choe succeeded. It required deftness of hand, meticulous attention to detail, some molten bronze, and a trough filled with sandy clay, but with his technique, one could mold characters, rearrange them to one's liking, and churn out as many exact copies as one wanted.

Joseon Dynasty scholar Hyon Song would describe Choe's process, still considered cutting-edge a hundred years later, this way:

"At first, one cuts letters in beech wood. One fills a trough level with fine sand of the reed-growing seashore. Woodcut letters are pressed into the sand, then the impressions become negative and form letters. At this step, placing one trough together with another, one pours the molten bronze down into an opening. The fluid flows in, filling these negative molds. . . . Lastly, one scrapes and files off the irregularities, and piles them up to be arranged."

Choe had invented metal movable type, and he did so a good hundred years before the birth of Johannes Gutenberg, who is widely credited with inventing the printing press and whose name, unlike Choe's, is the one that made the history books. It's just one example, in the view of many South Koreans, of how the country's contributions to science and culture have been roundly overlooked.

But as the Buddha once said, "A jug fills drop by drop." Just as Choe built upon the earlier technology of Chinese woodblocks, just as Gutenberg, evidence suggests, used Choe's invention to take printing a step further, scientists at Seoul National University would, by early 2003, pick up dog-cloning research where others—namely Westhusin's team at Texas A&M—had left off.

By then, Genetic Savings & Clone had severed its ties to Texas A&M and moved to Wisconsin, where its attempts to clone Missy continued without success. Texas A&M—having managed to create one embryo that briefly showed evidence of a heartbeat—had ceased its efforts to clone a dog.

On the other side of the globe, Seoul National University, operating independently and with no official connection to Missyplicity,

GS&C, or Hawthorne, stepped in, hoping to achieve the world's first dog cloning.

Learning from the past mistakes of others wasn't all the SNU scientists had going for them: Korean scientists, with fewer restrictions on the use of lab animals, were able to pursue their research more freely than their American counterparts. Animal welfare organizations were only starting to come into their own in South Korea, and in a country where dog meat is still served in some restaurants, there was little resistance to the idea of cloning canines.

All things considered, Korean culture served as a near-perfect medium in which to cultivate dog-cloning research: there was the stereotypical—but, in this case, apt—intensity of motivation to achieve. There was the equally stereotypical belief that Koreans, growing up using metal chopsticks, were more adept at "micromanipulation," the process required to take the nucleus out of egg cells and refill them with new DNA. There was a strong desire to be the first to accomplish the feat, fueled, in reality, more by national pride than the claims—valid as they proved to be—that cloning dogs could lead to medical advances for humans. And there was an abundance of dogs available for laboratory use, generally those that, lacking pedigrees, are still seen in Korea as lower-class canines, and most commonly end up on the streets or on the menu.

Just as researchers in the United States commonly visited slaughterhouses to obtain egg cells for their cattle-cloning experiments, Min Kyu Kim's first stop in the early stages of SNU's research was what's known as the dog market.

Kim, a postdoctoral student, was researching dog cloning for his thesis—before the SNU dog-cloning team was even appointed. While Genetic Savings & Clone was negotiating for eggs with spay/

neuter clinics in the United States, Kim was able to hop on the subway and get his egg cells from Moran Market, a block-long, open-air bazaar where customers can buy dog meat, or pick a live dog of no distinct breed (usually) from a cage and get it butchered to order.

When searching through discarded entrails yielded few viable eggs, Kim said, he made an arrangement with a dog farm, where dogs are raised for their meat. In exchange for veterinary services, the farmers allowed him to borrow dogs in heat, cut them open, harvest their egg cells, sew them back up, and return them to the farm.

The SNU project also had the advantage of receiving government funding—much of it aimed specifically at backing research that would lead to achievements that would bolster the country's reputation as a leader in science and technology.

One illustration of South Korea's perceived lack of worldwide respect can be found in the circumstances surrounding a book commonly referred to as *Jikji*. Printed in 1377, *Jikji* is a collection of Buddhist treatises and teachings compiled by a monk named Baegun to teach the Buddhist principles of self-denial and austerity, and the evils of excess. It is the oldest surviving book printed with Choe's metal movable type, and it predates by seventy-eight years Germany's Gutenberg Bible. Only one known copy remains, and it is in France. The volume was taken out of Korea by one Frenchman and sold to another, who bequeathed it to the National Library of France. It has been there since 1950, to the chagrin of many Koreans, some of whom would go on to launch a so far unsuccessful Bring Back *Jikji* movement.

That national inferiority complex, as some have dubbed it, and a concerted government effort to make South Korea a global leader in technology played large roles in the decision by scientists at Seoul

National University to pick up the ball that Texas A&M dropped and, for the first time, "bring back" a dog.

When the Missyplicity Project got under way at Texas A&M, cloning a dog was far from the mind of a soft-spoken Seoul National University professor named Byeong Chun Lee; nor did Woo Suk Hwang, his department head, view it as worth pursuing.

In the late 1990s, Lee had only recently, at the age of thirty, become a professor at Seoul National University, and at the time his work was focused on cloning cattle—a mission that, for him, was laden with irony.

Born in 1965, Lee was the youngest of seven children in his family—and as such was given the task of taking care of the family cow that lived on his father's modest farm. He attended high school in Cheunghu, and his family sacrificed heavily to send him to college, selling their entire rice crop to pay his first year's tuition at SNU. He was in his freshman year when his father died—killed by a raging cow. After that, Lee started to resent cows, as well as veterinary study, his intended field. He thought about other careers, or leaving school entirely. But his mentor at the university urged him to stick with it.

Lee's immediate superior at SNU, Hwang, was twelve years his senior and had also come from a modest farming background. Born in the aftermath of the Korean War, Hwang grew up in the central Korean province of South Chungcheong, raised by a mother who was widowed when he was five, and hard-pressed to provide for her six children. He eventually took a job at a farm, tending to cows, to help support the family.

"He was obsessed with cows," his mother told a *Washington Post* reporter in 2004. "There were times when he wouldn't come home for dinner because he was with those cows. Even during the nighttime, he would do his homework in the barn with the cows."

"They were my friends," said Hwang, who as a youngster imagined how much easier life would be for his family if there were a way to magically multiply cattle.

Hwang—the first member of his village to go past middle school—excelled in high school in Daejeon and won a full scholarship to Seoul National University. At SNU, despite being encouraged to become a medical doctor, he chose to be a veterinarian, spent a few years in a veterinary practice, then moved into research.

He returned to SNU as a teacher and researcher in 1987, the same year he—while under treatment for liver cancer—converted from Catholicism to Buddhism after visiting Jeondeung Temple in Incheon. The forty minutes of meditation he incorporated into his daily schedule after that was basically the only down time the hard-charging Hwang would have. He divorced his first wife, with whom he'd had two children, in 1988. In his first year at SNU, he took a small classroom, turned it into a laboratory, and named it the "Biological Control Research Center"—a place to pursue his dream of creating a genetically superior cow. Unsophisticated as the lab's equipment was at the time, Hwang and his researchers managed to produce a test-tube cow in 1995.

Lee was in his senior year at SNU when Hwang returned as a professor and researcher. After getting his master's and doctorate, and spending a year at Tokyo University, Lee, too, became a professor at SNU, taking over the position of his former mentor, becoming part of Hwang's research team, and overcoming his ill will toward cattle—a species most of their work was focused on at the time.

Lee was serious, studious, and soft-spoken. Hwang was charming, smooth-talking, and brash. Different as their personalities were, the two men shared one trait—intensity—and they were viewed as peas in a pod at the veterinary school, each exemplifying and demanding a stringent work ethic. Lee became known as "Little Hwang." He arrived at work at precisely 5:55 every morning, had his meeting with Hwang at 6:35, and devoted his research hours to cloning a cow. He honored Hwang's mantra—familiar to students and staff—that the days of the week were Monday, Tuesday, Wednesday, Thursday, Friday, Friday, and Friday.

In February 1999, Hwang and Lee cloned their first dairy cow—the fifth in the world. After that, Hwang announced the cloning of a genetically altered cow, resistant to mad cow disease. Handsome and smoothly articulate, Hwang traveled to other universities to deliver lectures, and spoke often of his hope to clone a Siberian tiger.

Dogs, at the time, were not part of the picture, though Lee says a veterinary client, as early as 1997, had asked him to clone her poodle. Benji, a female, had been diagnosed with pyometria, and Lee had removed part of the dog's uterus, right around the time the birth of Dolly the sheep was announced. "At that time, Benji's owner, she asked me, 'Professor Lee, can you possibly clone my dog?' I told her that would not be possible in the near future, but if it becomes possible, and if I clone a dog, it would be hers."

In 2001, Hwang sent Lee to the United States to learn about tiger cloning. That had yet to have been accomplished, but a Louisiana research institute, the Audubon Center for Research of Endangered Species, in New Orleans, was working on it. Using a frozen embryo, fertilized in vitro and implanted in a domestic cat, the center had produced Jazz, an African wildcat whose birth made headlines around the world. The mission of the center, which opened in 1996 and regularly collaborated with the University of New Orleans and Louisiana State University, was to use new reproductive technologies to reintroduce and ensure the survival of endangered species. One of its first projects, in conjunction with LSU, whose mascot is a tiger, was to try to save endangered tigers by using lions as surrogate mothers.

Through an arrangement with the Audubon Center and the University of New Orleans, Lee spent several months in the United States, researching tiger cloning in Louisiana. In 2002, he spent ten days researching pig cells as a visiting professor at the University of Chicago Medical Center.

While in Chicago, allowing himself some rare leisure time, Lee paid a visit to the Chicago Museum of Science and Industry. There his attention was drawn to an interactive exhibit about animal cloning.

"On the wall was a big screen [with pictures] of different animals. You pushed the button next to the animal's picture and it told you which ones have been cloned," Lee said. Though he knew all the answers already, he played along. First he pushed sheep, and got a yes. Pushing a button beneath a picture of a cow, he got another yes. Pushing a picture of playing children, he got a no. Pushing a picture of a cat and a dog, he got the answer "Yes/No," with an explanation: "Scientists have cloned a cat, but dogs have been extremely difficult to clone."

As Lee tells it, the exhibit served almost as a challenge. "I thought to myself, if I cloned this, maybe the world would be changed," said Lee, who still keeps a photo of himself standing in front of the display. Taken by a colleague, it shows a young Lee smiling at the camera as he pushes the glowing button with his middle finger, lighting up the words "Yes/No."

Upon his return to Seoul National University, Lee spoke to Hwang about getting funding to pursue cloning a dog. But Lee wasn't the only one making the suggestion. Taeyoung Shin, a key member of the Missyplicity team at Texas A&M, was also in touch with SNU researchers. Shin had completed his Ph.D. in Korea—with Hwang as his thesis advisor—before going to work at Texas A&M. As Texas A&M's dog-cloning efforts wound down, Shin suggested SNU pick up the research. Shin later left A&M to work for Genetic Savings & Clone when the company and university ended their relationship.

"In 1998, Dr. Shin went to Texas A&M for the dog-cloning project. He had many tries at cloning dogs, but they were not successful and they gave up dog cloning in 2002," said Min Kyu Kim. "I got so many ideas from Dr. Shin. We talked about it and he visited SNU in 2002 and we discussed dog cloning. He taught me about the manipulation of oocytes. He was very good at manipulation."

In August 2002, SNU's effort began, buttressed by information from Shin, and by the work Kim had done in the previous two years, culminating in his thesis, "In Vitro Oocyte Maturation and Somatic Cell Cloning in Dogs." Kim had found that egg cells needed to be enucleated, refilled, and implanted at a precise point in their maturation, and that trying to get them to mature in petri dishes wasn't a path worth following.

Kim says Lee approached him shortly after his return from the United States and suggested they work together to clone a dog. "He asked Dr. Hwang about getting the funding, and that's how the project got started," said Kim. "That's about the time all the team members started work, about five of us. We would work from four a.m. until eleven at night."

By then, Kim had already spent six months trying to get egg cells, in a petri dish, to mature to the point that they could be used in cloning. After giving that up, he'd spent another two years working on harvesting egg cells that were appropriately ripened to be right for cloning. Because dogs only go into estrus once or twice a year, and their ovaries only produce about 8 egg cells, compared with 150 in an ovulating pig, Kim, in his preliminary work, needed access to hundreds of them—and found it at a nearby farm.

Kim would monitor the dogs visually and through tests, waiting for a hormonal surge that he'd determined signified the optimum time was coming, four to seven days later, to harvest the eggs. Placing dogs under general anesthesia, he'd cut a slit in their undersides, find the opening to the oviduct, and flush the eggs into a petri dish. Under a microscope, he'd locate the unfertilized oocytes, then, with a tube in his mouth, he'd blow air into the medium to isolate the egg cells. He'd wash them, supplement them with fetal calf serum, enucleate them, wash them again, then introduce the donor cell through the same tiny slit through which he had removed their nuclei.

Then, with two jolts of electricity, he'd fuse the newly formed cells, and leave them in a culture for four hours, waiting for them to divide and form an embryo before implanting them into the oviducts of surrogate mother dogs.

He experimented with removing egg cells at different points in their maturation, with different methods of flushing them out, with different voltages in the fusion process, and with using egg cells from a different species, including those of cows.

By the time he was done, after more than two hundred harvestings, nearly five hundred mergings of the egg cells with skin cells obtained from the ear of the donor dogs—a Yorkshire terrier, golden retriever, and greyhound—and nearly fifty implantations of the resulting embryos into surrogate dogs, Kim achieved three pregnancies, the longest of which lasted thirty-three days.

The SNU team that would build on Kim's work was funded by the Korea Ministry of Education and Human Resources, through Brain Korea 21, a project established in the late 1990s "in response to concern over the relatively low standing of the nation's universities and researchers." The project's mission is to provide funding to researchers and graduate students whose discoveries could help drive economic growth in Korea in the twenty-first century.

While the scientists described the project as one that could lead to medical advances for humans, and while several promising avenues would later open up for that, most outside SNU's veterinary school, and a few who were inside, say the primary motivation of the researchers, at the time, was to garner a first.

"It was because nobody had done it yet. That made it worth pursuing," said Kim. "Also there was the reason that dog cloning could lead to medical advancements."

Making money by marketing pet cloning, while not a stated goal, was recognized as a possibility—at least in Kim's thesis.

"With the development of more advanced techniques, such as cloning and genetic engineering, the potential exists for a niche market, offering new and unique products," he wrote in his thesis. "Cloning and genetic modifications alone or in combination could be used to produce cats and dogs for use as models for biomedical researchers, service animals, or novelty pets."

Kim's thesis pointed to several planned commercial applications—including talk of "allergy-free" transgenic cats, the possibilities for marketing a fluorescent-green dog, and the increasing popularity of pet-tissue banks. He cited market surveys that reported that as many as one in four people would be interested in cloning their pet.

"As more pets are cloned, information will become available

regarding customer satisfaction," he wrote. "Do clones really look and act very similar to the original cell donor? Would they be just as happy with a different animal? Are all clones healthy and living normal lives? Answers to these questions are still unknowns, but will be key driving forces in the development of commercial cloning programs in the companion animal species. . . .

"The emotional value placed on pets will undoubtedly stimulate some pet owners to pay enormous amounts of money for a clone of their pet, especially once it is deceased."

While commercial possibilities loomed on the horizon, it was national pride, more than anything else, that led Hwang, Lee, and team to work nearly around the clock, their schedules subject to the ovulations of laboratory dogs. When Lee's family—a son, daughter, and wife—would complain about his long hours, their never having taken a summer vacation, their desire to go, for the first time, to an amusement park, Lee would ask for their patience, just a little longer.

"Please," he would say. "Wait until the dog is born."

14.

Cats, Clones, Clowns, and Con Men

■ ■ ■

One of the most striking differences between a cat and a lie is that the cat has only nine lives.

—*Mark Twain*

MADISON, WISCONSIN

February 14, 2004

A year to the day after the death of Dolly the sheep, two years to the day after the announcement of the birth of CC, the world's first cat clone, Genetic Savings & Clone launched the "First Nine Lives Extravaganza"—a plan to clone nine cats: three for friends and family, six for the general public. The announcement came on Valentine's Day, as had Hawthorne's announcement, four years earlier, about establishing Genetic Savings & Clone.

The first cat cloned by the company was Tahini, who Hawthorne said belonged to his son, Skye. The two clones of Tahini, born in the summer of 2004, and named Tabouleh and Baba Ganoush, were the spitting image of their donor. In announcing their births, Hawthorne got in an apparent dig at the Texas A&M scientists with whom Genetic Savings & Clone had parted ways after they produced CC, a calico cat clone that bore little resemblance to its donor.

"These two remarkable kittens should finally put to rest the issue of resemblance between clones and their genetic donors," Hawthorne said on the website. "When performed by a skilled team using sufficiently advanced technology, clones resemble their donors to an uncanny degree—just as we predicted."

Another company prediction—that the cloning of dogs was near—had yet to be realized, much to the disappointment of the man funding the research, billionaire John Sperling. But having conquered copying cats, Genetic Savings & Clone turned its focus to that market. On the company's website, Hawthorne guaranteed healthy cat clones, and full refunds to any customers who were not satisfied. He promised to keep the identity of clients anonymous if they so requested. Customers, once their cat clones were born, would receive an all-expenses-paid trip to Sausalito to attend a party at which the clones would be presented, and a video on the cloning of their cat, including the moment of birth—all included in the $50,000 price.

In August 2004, with a third cloning success under their belt—a cat named Peaches—Genetic Savings & Clone began taking the clones to cat shows around the country. In October, Tabouleh and Baba Ganoush drew a crowd at the New York Cat Show in Madison Square Garden. A few months later, in December 2004, GS&C presented the first cloned cat to a paying customer.

The kitten's name was Little Nicky, and she set off a big uproar—at least compared with that which pet cloning had caused until then. The Missyplicity Project, the cloning of CC, and the taking of cat-cloning orders had led to some criticism from the animal welfare community. But the exhibition of the animals, and news of Nicky—the first cloned household pet to be sold—triggered a wave of disapproval from some bioethicists and such organizations as the American Anti-Vivisection Society.

"At the cat show, everybody was saying, 'Oh look, how cute, cloned cats.' But they weren't seeing the dirty stuff behind the curtain. The same with Little Nicky—it was just the adorable kitten and the happy woman," says Sue Leary, president of the American Anti-Vivisection Society. "Behind that cute picture of a kitten was a lot of bad stuff, and this wasn't just a cute, clever, high-tech new way of breeding. This is fraught with problems, and lots of animals are behind that curtain—hundreds are subjected to painful and invasive procedures just to produce a single cloned cat or dog. We thought there was a risk that, even if it didn't turn out to be a billion-dollar business, pet cloning could gain public acceptance—unless somebody spoke up about what was really going on."

In addition to addressing animal welfare concerns, AAVS has portrayed the dog-cloning business as coming dangerously close to a scam. "We point to the consumer-fraud issue, because you're not getting the same animal, and clearly—even though the companies might have eventually put a disclaimer someplace—clearly it was being sold on the impression that you were going to have the same animal back."

Genetic Savings & Clone was, at least, actually pursuing cloning in the lab—and, in the case of cats, accomplishing it. But some of the other companies that were offering the service were not involved in research. They were merely banking DNA, if even that.

When it came to Clonapet, the arm of Clonaid aimed at "wealthy individuals who wish to see their lost pet brought back to life," there hadn't been a peep—no announcement of a dog, cat, or any animal being cloned. Raël, the religious leader who established the company, had stepped back from the business side and was once again pursuing auto racing. In one race, Raël finished in fourth place, behind actor Paul Newman, after which the two men exchanged gifts. Raël gave the actor a copy of his book *Let's Welcome Our Fathers from Space*; Newman responded with a bottle of his salad dressing.

Raël had appeared before the House Energy and Commerce Subcommittee when it held hearings on human cloning in March 2001, with his hair tied in its customary topknot, a large medallion on his chest, and clad in his trademark white-satin, shoulder-padded jumpsuit. (The pro-cloning preacher once told a reporter he made all his own clothes, because "I am against uniformity in all things.") Raël carried with him a manifesto signed by thirty-one scientists and thinkers in support of human cloning—from Francis Crick, co-discoverer of the structure of DNA, to author Kurt Vonnegut. He told lawmakers to get out of the way of science. "Stopping science," he said, "is a crime against humanity." Cloning needed to be legalized, he added, to "protect the rights of the unreborn."

The next year, just after Christmas, company leader Brigitte Boisselier announced the birth of a baby girl—cloned, she claimed, from the skin cell of a thirty-one-year-old member of the Raëlian sect whose husband was infertile. The announcement was met with skepticism. Clonaid never managed to clone a pet, and some thought they were doing nothing more than capitalizing on the grief of the infertile and bereaved.

If so, they weren't alone. Cloning, since not long after Dolly the sheep was born, has had a way of drawing in bad actors, rogue scientists, eccentric characters, and businessmen of questionable repute, who have made offers they couldn't back up, raised hopes only to dash them, and blurred the ever fuzzier line between science and science fiction.

"Why is Congress debating this by talking to someone who says he flies around in flying saucers?" Michael West, the CEO of Advanced Cell Technology, the leading genetic engineering firm in the United States, said of Raël's appearance at the congressional hearing.

By 2004, another player entered the pet-cloning market, one whose background contained no flying saucers—just hot-air balloons.

When a company called ForeverPet was launched in the United States that year, promising cut-rate cat cloning—or at least less than what Genetic Savings & Clone was offering at the time—most press accounts failed to mention that the man behind it had no expertise, no laboratory, and more experience with cells of the prison variety.

Simon Brodie had come to the United States from London, via Toronto. In 1991, in Uckfield, England—four hundred miles south of where Dolly was cloned—Brodie started a company called Cloudhoppers, which offered hot-air balloon rides. A year later, he tried to sell it to an investment firm for £4 million. That summer, an investigation by the *Evening Argus* newspaper revealed that Brodie's claims about the company's worth, and its sale, were exaggerated. It consisted of only two balloons. Employees, claiming they hadn't been paid, started leaving, including his chief balloon pilot. The Civil Aviation Authority suspended the company's license. The telephones were cut off, Brodie's Lamborghini was repossessed, and the company was evicted from its building.

In August 1992, Brodie was charged with false accounting and eight counts of theft involving money swindled from banks and gained through double-charging customers' credit cards—all in connection with Cloudhoppers. In 1994, he was sentenced to two and a half years in prison.

Upon his release, Brodie opened a couple of short-lived computer-related ventures, moved to Toronto, and then resurfaced in California, having formed a new company, Geneticas Life Sciences, which, through its divisions, seemed to focus on various ways to make animals more human-friendly. One division, Genetiate, pushed "NightSave Deer," a project that would create fluorescent deer by implanting jellyfish genes in them. It was aimed at reducing the number of car-damaging highway collisions. ForeverPet, formed in August 2004, offered pet cloning, at least on the Internet, until 2006. It never cloned an animal. Still, it portrayed itself as being "at the forefront of cloning research and development."

According to a press release, ForeverPet planned to charge $19,950 for a cat cloning, with additional clones of the same donor animal priced at $4,995 each—"close to a third of the competition's rate," Brodie boasted. Five months later—though he said cloning orders were pouring in—Brodie was headed in a new direction. In November 2006, he announced that another division of Geneticas—Allerca, Inc.—planned to produce hypoallergenic cats. The cats would be created, through cloning, with the gene that produces the most common feline allergen—a protein called Fel d 1—effectively silenced in the process. He predicted sales of 200,000 genetically engineered cats a year.

Those plans were dashed when David Avner, an emergency room physician in Lone Tree, Colorado, went to court claiming Brodie had stolen the business plan the two men had arranged to pursue together. In his lawsuit, Avner said Brodie suddenly pulled out of the deal and, weeks later, was incorporating the company on his own in California.

The judge ruled in Avner's favor, enjoining Brodie from developing and marketing cloned hypoallergenic cats.

In June 2006, Brodie was back with another press release, announcing hypoallergenic cats would soon be available from his company—created not through cloning but by selectively breeding cats in which the allergy gene had naturally mutated. *Time* magazine honored the discovery as one of the "Best Inventions of 2006."

Lifestyle Pets marketed three "brands" of cats, and in 2008 they added a hypoallergenic dog to their inventory—the Jabari GD, which sells for $8,950. During President Obama's quest for a hypoallergenic dog for his family, Brodie was among those who contacted the White House. The Jabari joined one other dog included in the Lifestyle Pets product line: the Titan Protector Ultra, which, despite a name that sounds like a brand of condom and despite a price tag of $120,000, was basically a trained German shepherd. "While the price may seem high, this is not usually an issue for our clients. . . . After all, what price can you put on your family's safety?" Brodie's website asked.

Like most of Brodie's ventures, the Titan was aimed at the very rich. Like most of his ventures, the Titan was something fairly ordinary marketed as rare, exotic, and exclusive. And like most, it was aimed at using animals to make life more comfortable and convenient for humans—whether it was by making deer glow in the dark or creating dogs and cats that wouldn't make us sneeze.

An even more glaring example of that mind-set—the view that animals and, more specifically, pets can be adjusted to our liking—was FlexPetz, a "pet rental" company formed by Brodie in 2007, with outlets in New York, San Diego, and Los Angeles. Members of FlexPetz could "spend from just a few hours to a number of days with each of our dogs." Dogs, the website added, were available in varied sizes to "ensure compatibility with our members' individual lifestyles and unique circumstances."

For a $100 fee, and another $50 a month, one could become a member of FlexPetz and rent a dog in what the company promised would be an expanding number of major cities worldwide, thereby enjoying all of the advantages of having a pet—whether it be for friendship or as a fashion accessory, thug-repellent or chick-magnet—and none of the responsibility.

The initial reaction to the concept, at least among news media, was—as is often the case with dog-related stories—fawning, pun-laden, and superficial, and failed to delve into either Brodie's background or the ethical questions the business raised. FlexPetz opened three branches, and had two more in the works, in London and Boston. But opposition was mounting from animal welfare organizations and others who pointed out that rental dogs, however much they may please humans, would themselves be scarred by the experience. Boston's city council passed a law banning rental pets, and not long after that FlexPetz went on hiatus.

While pets on demand, as a business concept, didn't gain a foothold, the idea of pets for perpetuity—through cloning—was making major strides in 2004, and garnering more criticism, as well.

In the wake of Little Nicky, and a second commercially cloned cat—Little Gizmo, delivered to a Southern California man on Valentine's Day 2005—the American Anti-Vivisection Society filed a petition with United States Department of Agriculture, asking it to regulate facilities that were going to clone or genetically modify animals. The USDA didn't go that far, but it did reaffirm that those who exhibit cloned animals publicly open themselves up to regulation and inspections. Working with the Sacramento-based organization United Animal Nations, the society also backed a bill to outlaw pet cloning in California in 2005. The bill didn't pass, but it did help bring slightly more serious attention to the issue.

Amid the growing controversy, the owner of Little Nicky, the first cat clone created strictly for commercial purposes, opted to keep her identity a secret. Never publicly identified by more than her first name, Julie was described in news reports as a satisfied—thrilled, even—customer. She spoke of the uncanny resemblance Little Nicky had to Nicky, and after accepting the cat at a party Hawthorne threw in Sausalito, she returned to Texas with her privacy intact.

In the years that followed, Little Nicky developed undisclosed medical problems, Julie hired legal representation, and, while still cloaking her last name, she went online to leave Internet posts critical of Hawthorne and Genetic Savings & Clone—in response to an article on Lou Hawthorne in *New Scientist* magazine.

In a July 3, 2008, comment signed "JulieandNicky"—confirmed by sources to have come from the Texas woman—she complained that Little Nicky's health was poor, and that the cat was never independently verified to be a clone. In addition, she wrote: "This Nicky has a medical condition that the first Nicky never did. GS&C has never followed up on Nicky's health (even after stating this in a biotech debate). I have noticed that Mr. Hawthorne never mentions the cats he cloned at GS&C. Are they healthy, has his company followed them for possible cloning-related medical issues?"

Hawthorne posted a comment in response the next day: "All of the cat clones we produced at GS&C are, to my knowledge, healthy. . . . Of the 7 we produced, I know the whereabouts and well-being of 3 (not counting Little Nicky). . . . Regarding Little Nicky's health, I'd be interested in learning more. I suspect that if Little Nicky has any health issue, it's an acquired one, not a genetic one."

Hawthorne went on to point out that, while his company's failure to provide certification that Nicky was a clone was an "unfortunate oversight," he had repeatedly offered to pay for tests by an independent laboratory if Julie would provide a DNA sample of the cat.

"After Little Nicky was delivered—to an overjoyed client, by the way—our ability to certify the clone depended on the owner's willingness to provide a DNA sample. . . . Rather than accepting our offer though, she sent us email after email complaining about our company and our staff. This exchange went on for over a year and involved several hundred emails and phone calls total. Literally DOZENS of times, we offered to arrange DNA certification for Little Nicky."

A week later, "JulieandNicky" responded to Hawthorne's response on the *New Scientist* website: "Your version of what transpired is quite an interesting tale. May I remind you, it was you that went in front of the national media and proclaimed Nicky was a clone. Certainly you must have the test results to back up your claim. . . . Might I also remind you that I contacted the DA's office in Marin County and requested that GSC take part in voluntary mediation to settle this and you refused. . . . For $50,000, Genetic Savings and Clone certainly should have provided us with documentation that Nicky is indeed a clone. . . . Quite sadly, Nicky is not a healthy cat. You have absolutely no knowledge of Nicky's health issues, so I don't know how you can . . . say his health issues are acquired and not genetic."

Hawthorne found a much more satisfied, much less media-shy customer in Liam Lynch, a Los Angeles–area composer, musician, puppeteer, director, and writer.

In 2002, Lynch had purchased a cat from a shopping mall pet

store; he named him Frankie Forcefield. Lynch owned other cats, some of them far more exotic and expensive than Frankie. But it was Frankie—and his playful quirkiness—that captured Lynch's heart and mind and got him thinking a descendant might be nice. Traditional offspring were out of the picture, because Frankie was neutered. Having read about advances in pet cloning, Lynch had considered banking Frankie's genes, but he hadn't gotten around to it.

Four years after he bought Frankie, in May 2006, the cat was struck by a car and killed. Lynch didn't hesitate. He wrapped the cat's corpse in plastic trash bags, stuck it in his refrigerator, and called Genetic Savings & Clone, which by then had relocated to Wisconsin. As per the instructions he received, he contacted his vet, who took nine tissue samples from Frankie, placed them in the company-supplied "Biobox," and sent them overnight to the company's lab.

To Lynch, who was writing a screenplay at the time, it all seemed like science fiction coming true. Lynch was behind the creation of MTV's *Sifl and Olly Show*. He directed music videos, and had cowritten the original music in the movie *School of Rock*, composed the music for the MTV animated series *Clone High*, and directed the movie *Tenacious D in the Pick of Destiny*. He's also the creator of *Lynchland*, an offbeat variety show he produces for online consumption, and on which he kept viewers abreast of his quest to have his cat and "best friend ever" cloned.

"While he was alive, I started looking into banking his genes," Lynch tells viewers in one episode. "I used to joke with friends that I would have him cloned and have a bunch of copies of him, at various ages, walking around my house."

Lynch told viewers Frankie's cells were multiplying in a lab as he spoke: "They take these little pieces of skin back to the lab and it takes two to three weeks to see if anything's going to happen. This is completely up to nature, no science can make this occur. . . . It's new life growing, new cells, kind of resparking. Frank did grow new cells. He grew about 8 million new cells. So in a way Frankie Forcefield is alive, but he's in cellular form. He's in his most simple form. . . . I sent these off thinking it is up to nature, God, Buddha, or Frank to decide if life comes back in these. There was about a 20 to 30 percent chance. . . . Genetic Savings & Clone has contacted me and said it's working. . . .

"I'm not going to call it 'Frankie Two,' or, as my dad suggested, 'Frankie-stein.' I'm going to call him Finnegan Forcefield, and Finnegan Forcefield is alive right now in cellular form. . . . Will Frankie's clone Finnegan like all the treats Frankie liked? Will he want to sleep in the same spot? . . . Who knows? But it is serious cutting-edge technology. . . . The future is happening and it's just totally wild and I love it."

In September 2006, Finnegan Forcefield was delivered, looking and acting just like Frankie, says Lynch. While Lynch doesn't think he has the same cat back, he does feel he has a new friend, and one that serves as quite the conversation piece. His friends think it's awesome. As for the critics, he doesn't care.

"I'll let people think what they want," he said in an interview. "I know some people have moral issues with it. They feel it's playing God. I usually tell them to watch a dog show on TV and remind themselves that all those extremely different breeds were wolves until we started breeding them into corners. It's just the next step in something we've already been doing for hundreds of years."

Lynch was Genetic Savings & Clone's last customer.

After terminating its arrangement with Texas A&M, GS&C had

moved from Texas and, in 2005, opened a new lab in Waunakee, a suburb of Madison, Wisconsin—an area that had just seen a company that cloned dairy cows, Infigen, shut down. GS&C hired some former Infigen employees and opened up in Suite J of the Waunakee Business Center, with plans for a grand opening on Valentine's Day.

But the company was losing money—especially after it lowered the cat-cloning price from $50,000 to $32,000 in an attempt to draw more customers. The losses—coupled with that which it was spending on still trying to clone a dog—added up. The facility, the process, and the technical staff, which by then included the scientist on the team that cloned the gaur, were proving too costly.

"We were losing money on every cat," Hawthorne says. "The media widely misreported when we shut down that GS&C closed for lack of demand. That wasn't the case at all." He says he signed ten refund checks for customers who had already paid for future clonings. "I decided to take it dark, lay off everybody we don't need, figure out what we needed to accomplish, and not be a company that overcommits. We promised to be this big cloning company that was going to do all these animals. We promised low-price cloning in the future. We just bit off more than we could chew."

In September 2006, Genetic Savings & Clone sent letters to clients whose pets' tissue they were storing—about a thousand of them. Those who'd placed their orders for cloning cats would get refunds. Those awaiting the cloning of dogs would have to keep waiting.

The letter informed customers how to go about transferring the tissue of their beloved pets to gene banks elsewhere, including ViaGen, a gene-banking and livestock-cloning outfit also connected, at the time, to Sperling's empire. Most of the millions of living cells, harvested from pet cats and dogs, before and after death, would be moved to new gene banks, allowing the hopes of their owners to live on.

From all appearances, GS&C was dead and, with it, the hopes of hundreds of customers who had banked their dogs' tissue for the day dog cloning became a reality—among them, James Symington and Bernann McKinney.

But like Frankie Forcefield, like Tahini and Nicky, and like Chance the bull, Genetic Savings & Clone—under a different name—would be back.

15.

Second Chance

■ ■ ■

As a dog returneth to his vomit, so a fool returneth to his folly.

—*Proverbs 26:11*

LA GRANGE, TEXAS
March 2005

Second Chance was getting one.

Ralph Fisher, after his cloned bull Second Chance attacked him in 2003, dislocating his shoulder, let bygones be bygones. He was a little more careful around the animal, and hadn't worked him fully into the rotation for Ralph Fisher's Photo Animals. But he chalked up the earlier unpleasantness to the bull's youth, still convinced that Second Chance wasn't just the spitting image of the original but had the same peaceful soul, too.

Publicity about the cloned Brahman bull—which didn't hurt Fisher's company—had died down in the years after his birth in

1999. Second Chance made a handful of public appearances, usually at $1,000 a shot, but always as the cloned bull, never as the gentle one. At each, because of his earlier attack on Fisher, he was kept in a pen and tied to a tree. Fisher, though convinced Second Chance was a gentle beast, wanted to be sure before he started letting toddlers climb aboard for photos.

For the most part, Second Chance lived a quiet, petlike existence, mostly hanging around outside the Fishers' kitchen window, as opposed to in the pasture with the other livestock, wandering the yard or settling down to snooze under his—and his predecessor's—favorite tree.

In early 2005, Second Chance almost appeared on David Letterman's show, as his predecessor had done in the 1990s. It would have been something of a television first, with the show being live but the animal being a repeat. As it turned out, the show had been overbooked, and its staff called the Fishers to cancel. While that plan fell flat, Second Chance did make it on TV two years later, after a crew from *This American Life* came to town to do a report on the cloned bull. The popular and critically acclaimed Chicago Public Radio program, hosted by Ira Glass, was negotiating with Showtime to produce a TV version. Second Chance was to be part of its pilot episode.

The crew planned to stay for three days to film around the ranch and interview the Fishers. Ralph, folksy and personable on camera, explained what a sweetheart the original Chance was. "He was just like a big bundle of whatever. . . . Like your favorite dog, he'd lick your face." When Glass asked him if having a cloned version of his favorite animal around was painful, if it served as a reminder that the first one was gone, Fisher quickly straightened him out.

"No sir, it's just the opposite, you're just exactly as wrong as you can be. . . . So far right now I think we got about 95 percent of him back—the same qualities, the same fun. That satisfies me. That's better than zero," Fisher said. "When he [the original] was lying out in that pasture [dead], he was a zero. . . . We wouldn't ever have had any enjoyment out of him anymore. There's a tremendous difference between zero and 95 percent. We were allowed to have most of our feelings about him back."

When Glass questioned Sandra about having a bull for a pet—"they seem so much less demonstrative," he said—she interrupted. "Wrong. That's wrong. It really is like having a pet dog or cat, except the size." The Fishers spoke of how both Chance and Second Chance loved to nuzzle up against them, to be brushed and petted, and how they both soulfully stared, just like a dog or cat might, until their meal arrived.

Second Chance, like the original Chance, didn't seem to mind all the cameras and attention. "He'd been real good. We'd saddled him up, and we'd been sitting on him and petting him all day," Fisher says. But between all the filming and interviewing, the bull's feeding time was delayed an hour or two.

The Fishers were eager to show Glass and his crew how Second Chance ate, because it was the exact same way Chance used to eat: taking a mouthful of feed from the bucket, then—contrary to the manner of most bulls—pulling his head out to chew and swallow. "It's real unique," Sandra says.

Some members of the crew, including a cameraman, accompanied the Fishers to the barn. Ralph filled Second Chance's feeding bucket, and the bull started eating.

"They were trying to get that on film, and the cameraman asked if I could move him over a step because he wasn't right under the spotlight. So I was just moving the bull a little bit," Fisher recalls. He pushed on Second Chance's forehead, trying to make him take a step back. Then he pushed again.

The third time, Second Chance pushed back.

"I was standing right across the fence watching and Ralph had the feed bucket in his right hand and he pushed on Second Chance's forehead, and then he pushed him again," Sandra says. "And when he pushed him a third time . . . all of a sudden, Ralph is doing a somersault up in the air."

He hit the ground with a thud, and the bull came at him.

"I did kick him," Ralph Fisher says. "I remember I was on my back kicking. You can't just let them have at you. Everything happened so quick, probably two or three seconds and the whole thing was over."

"You were on your back kicking him in the face," Sandra interjects, "and I was slapping Second Chance on the hump. I was yelling 'Stop it!' because he was going after him. When Ralph's on the ground he's not the master. But he was able to push himself out of the way, and as soon as he got up, Second Chance turned and started eating the feed Ralph had just dropped."

Ralph, after getting up, looked down and saw a puddle of blood. He peered inside the gash in the crotch of his pants. "My left testicle was just hanging down, extended on a cord. I just sort of packed it all back in and we went to the hospital."

At the hospital, Ralph received eighty stitches and was kept overnight.

"I was just fortunate," he says. "I'm blessed. I'm lucky. He should have killed me right then."

"Ralph," Sandra cuts in. "Go get your pants."

Ralph returns to the living room and holds up a pair of blue jeans with a four-inch gash in the crotch. "These were brand-new Wranglers," he says. "And look at these underwear." He holds up a pair of briefs similarly ripped. "They were practically new double briefs. He hit me with such force that his horn went through the jeans, through the underwear, and tore my scrotum."

The ripped Wranglers are just one exhibit in the Fishers' museum-like home, at the end of a long, dusty driveway outside La Grange. They've amassed a large amount of bull memorabilia—from the shoulder mounts on the wall to the original Chance's pelt, kept in a box in a closet, to numerous scrapbooks, news clippings, photos, and, of course, the video of *This American Life*.

"You wanna see it now?" Ralph asked after getting the VCR hooked up. The television version aired in 2007, more than a year after the story ran on the radio broadcast. The episode, called "Reality Check," contains three stories about people pursuing their dreams, only to get snapped back to reality.

In a taped interview, Mark Westhusin, the scientist who cloned Chance, talked about the flood of pet cloning requests he'd received since news came out about the Missyplicity Project. He pointed out, yet again, that cloning does not bring the same animal back. "It's not resurrection; it's reproduction." But Ralph, in an interview from his hospital bed after the attack, made it clear that—despite the second attack—he was still not convinced of that.

"This is exactly the same type incident as the first one. . . . I'm just thinking that we just have to have a lot of faith in things to work out. I forgive him. I just shouldn't have been that close." He noted, as he noted after the first time Second Chance attacked him, that he didn't get the original Chance until the animal was seven years old, and that maybe Second Chance—not quite six at the time of filming—was still growing into that mold.

What would convince him otherwise? "After he's seven, then he has to attack me. . . . I'm going to walk out of here tomorrow or the next day and I'll walk out there and give him another bucket of feed. That's what you have to do."

At the end of the segment, Ralph turns off the TV and Sandra brushes her face with her hand. "I cry every time I see it," she says.

Ralph doesn't place any blame on the TV crew for the mishap or, for that matter, on Second Chance. Instead, once again, he blamed himself.

"I shouldn't have pushed on him. I actually pushed on his head. I just had to push him back about five feet, so instead of getting the lead rope and walking him around in a circle and putting him where he needed to be, I thought I was going to just push him back. And Lord, by the third time I thought, 'I shouldn't be doing this,' and sure enough, he came straight up and that's when he did all that. It was just circumstances, it was just all of a sudden. That's the beauty of Brahmans. They're like a light switch. That's why they use them as bucking bulls in rodeos. They get easily aggravated."

After Second Chance turned seven, Ralph says, the bull did seem to mellow out. There were no more sneak attacks, but then again, Ralph was being more careful. "He just seemed to be getting more gentle. He was just getting there."

By then, though, the stomach problems that plagued Second Chance all his life were turning more severe.

Second Chance was diagnosed with "ruminal outflow disorder," meaning that what was going into the bull was not coming out. In one of the two surgeries Second Chance underwent for the problem, veterinarians found and removed a fifty-five-pound clump of undigested feed.

"He kept getting clogged up," Ralph explains. "There's a nerve that makes the muscle react and pushes the food through the system—through all four stomachs. But he had some nerve damage. They don't know what caused it. Nothing traumatic happened to him; it's just how he was. They couldn't fix the nerve problem, therefore the muscle wouldn't work. He just couldn't digest his food and he'd bloat."

After he turned eight, the outlook was bleak.

"They had said that the clones were going to outlive the donors," Sandra says. "Chance was nineteen when he died. I guess we just thought we'd die before Second Chance did, and I feel kind of stupid now that we were so naive. It just never occurred to us that he was going to die. After the last visit, they basically said they were going to send him home to die. He was there six weeks, and we were pretty disappointed that Westhusin and the cloning people didn't even go look at him."

"They had cloned something like seven other species after they cloned him—cat, pig, deer, horse, and all that other stuff. . . . I guess in a way he was old news," Ralph says.

Back home, Fisher kept Second Chance in the barn for several days, to restrict his food intake. "Then we figured if he's going to die anyway—we just as well oughta let him get out and eat whatever he wants and enjoy himself."

The Fishers found him dead on March 11, 2008.

"We actually went down there that morning to take some final photos," Sandra says. "I think he had just died. I don't know whether he had a heart attack or what. It was at that point, when he was lying there dead, that he really looked more like Chance than ever before."

While more than two hundred news organizations had shown

up for the press conference about his birth, Second Chance's death was barely acknowledged in the media—and then only locally. No autopsy was performed, so there is no documentation on whether being a clone shortened his life.

"That's the first thing people ask," Ralph says. "The scientists say there's probably no correlation, but they don't really know. And I'm not real sure. . . . They may not absolutely make a dishonest statement, but they could possibly go around an issue to not have to talk about it. A lot of things, if you ask them, they just kind of beat around the bush, because they don't know, either."

The Fishers buried Second Chance on the ranch, but kept his horns and pelt to have a shoulder mount made, which they plan to hang on a wall along with that of their deceased Texas longhorn, Tumbleweed.

"I wish I could have afforded to clone the clone," says Fisher, who was inducted in the Texas Rodeo Cowboy Hall of Fame in the spring of 2009. "I hate that he died, because he was just getting more mellow and minding his own business. I thought he was almost there. But we'll never know. We'll never know."

Fisher says he has, finally, come to agree with Westhusin—that Second Chance and Chance, despite their similarities, were different animals.

"Chance was one of those kinds of pets that are so rare—completely trustworthy with kids and grandkids. That's why it was so hard for us to get it through our heads that Second Chance was not the same animal. It took me years, and a couple of hospital stays, to figure it out. Westhusin was right. They looked the same, but there's something going on in that brain that's not the same. We learn by environment and life experiences and there's something that the old guy got that this one didn't get."

16.

Snuppy

■ ■ ■

> What is a dog, anyway? Simply an antidote for an inferiority complex.
>
> —*W. C. Fields*

SEOUL, SOUTH KOREA

April 24, 2005

Though a genetic duplicate, Snuppy was, arguably, unique.

He was obviously, though sprung from the womb of a yellow Lab, a purebred Afghan hound. He was seemingly one country's resounding achievement—South Korea's. And while it would be debated whether he was a legitimate clone, he was clearly a first.

It took three dogs—not counting hundreds more whose eggs were harvested and wombs were borrowed—to make him.

The first puppy clone in the world was derived from the skin cell of an American-born Afghan hound with an Indian name. His cells were fused with the egg cells of a Korean farm mutt with no name.

And he was delivered from the womb of a yellow Lab named Simba, the Swahili word for "lion."

More than two hundred reporters attended the August 3, 2005, press conference at Seoul National University announcing, after three months of secrecy, the April 24 birth of Snuppy, whose name is a merging of the university's acronym, SNU, and the word "puppy." Snuppy; his donor dog, an Afghan hound named Tai; and Simba, the surrogate mom, named after the *Lion King* character, were all trotted out for the press.

The vast supporting cast—of dogs and humans—got no curtain call: not the 120 other bitches in whom implanted embryos failed to form into puppies; nor the more than 100 "farm dogs" from whom eggs were surgically harvested; nor the fetus who was stillborn or the newborn pup who lingered twenty-two days, much of that time struggling to breathe, before dying; nor the scientists at Texas A&M, who spent five years laying the groundwork for canine cloning. This day was about Snuppy, and the South Korean scientists who created him, and the man who, at least on paper, was in charge—Woo Suk Hwang.

"It's a miracle. We thought it would take five years," a jubilant Hwang told reporters.

To accomplish the feat, SNU researchers—according to their numbers—had extracted eggs from about 115 dogs. They removed the nucleus of each, delicately squeezing it out through a small slit made in the cell's outer membrane. Using skin cells from the ears of Tai, they inserted an entire cell into each emptied egg, through the same small opening. They applied an electric charge, in two pulses, to each egg, fusing it with the new cell inside it, then immersed each reconstructed cell in a chemical bath until they started to divide.

About 75 percent of the attempts resulted in new cell division, leading to embryos. The 1,095 reconstructed canine embryos were transferred into the oviducts of 123 more dogs through surgery—about 9 per surrogate.

Three pregnancies, in three surrogate dogs, resulted. Snuppy was delivered by cesarean section sixty days after embryo implantation, weighing in at one pound, three ounces. The second dog born, named simply NT-2 (for nuclear transfer), experienced respiratory distress during his first week and died of pneumonia after three weeks.

Those messy details didn't rate mention at the unveiling. All eyes were on the sleek new pup, who—like Dolly before him—seemed to bask in the attention. He posed cooperatively for photos with his surrogate mom, his donor dog, and the man who, though he shared credit with the entire team, was widely viewed as his creator, Dr. Hwang.

Hwang was on a roll by then. Having branched into work on human cell cloning, the doctor of veterinary medicine had announced just two months earlier that he had cloned eleven human embryonic stem-cell lines, a follow-up to his reporting the previous year that he had created the world's first cloned human embryo.

Since then, he'd been enjoying the kind of adoration generally reserved for rock stars and sports legends. Those reported achievements, and the potential they held for treating and curing human disease, had catapulted him to international fame, and to hero status in South Korea, where the government bestowed upon him the title of "supreme scientist." He was regularly stopped for autographs and greeted with spontaneous outbursts of group applause.

By 2005, Hwang—having bested his own liver cancer, having created cloned cows and pigs, and having reportedly cloned human embryos and derived stem-cell lines from them—was enjoying his own brand of immortality, at least in South Korea. Snuppy was icing on the cake, which, by the way, is the cloned dog's favorite food.

With Snuppy's birth, dog logged in at No. 18 on the list of species cloned—the most difficult animal yet to genetically duplicate, the most expensive, and the closest, at least emotionally, to man.

The cloning of man's best friend made the front page of *The New York Times*, the *Los Angeles Times*, *The Wall Street Journal*, and *USA Today*, and it would go on to be named the "most amazing invention of the year" by *Time* magazine, on whose cover a blissfully oblivious Snuppy lay, his silky white front paws flopping across the magazine's trademark red border.

Most media reports—once they got beyond the cute factor—moved on to outline the medical advances for humans to which the SNU scientists said Snuppy could lead, and, to a lesser extent, the commercial applications of pet cloning, which the scientists went out of their way to downplay.

"We are not in the business of cloning pets," said Gerald Schatten, the only American member of the SNU research team. "We transferred 1,095 embryos and got one dog. If I were an investor and someone came in and said we have a technique that works at 0.09 percent, I'd say that doesn't sound like a good investment."

The journal *Nature*, which published the Snuppy research paper that month—Byeong Chun Lee's name appeared on it first; Min Kyu Kim, though he had been promised the No. 1 position, agreed to be listed second—characterized the development in an editorial as having "some scientific significance," but noted the process was "remarkably inefficient."

"This offers scant prospects for commercial pet cloning. . . . It is unlikely that even the most obsessive pet owner would contemplate preparing more than 100 failed pregnancies for just one successful birth—especially when there is no guarantee that the cloned dog will behave like the one they hope to duplicate. In such circumstances, the cloning of dogs for pet owners remains ethically indefensible."

Seoul National University's veterinary school had spent about $1 million on the project—far less than either Texas A&M or Genetic Savings & Clone, the private company that, once it severed its ties with A&M, continued its attempts to clone a dog at a new lab in Wisconsin. By then, it had spent close to $20 million.

Genetic Savings & Clone's public relations man at the time, Ben Carlson, was able to come up with a positive spin: "This justifies our investment in the field," he said. "We suspected that if anyone beat us, it would be the Koreans." Carlson said GS&C expected to produce a cloned dog by the end of the year. It didn't, and shut down in 2006.

That the Korean scientists were able to do, officially, in less than three years, what U.S. scientists had spent seven years trying to accomplish was partly a result of their level of commitment, their round-the-clock, no-days-off work ethic; partly a result of their pinpointing, through test after test, the proper time to harvest the egg cells; and partly a result of their easy access to dogs, obtainable by the hundreds, with few worries about animal welfare regulations.

"The truth is they used the exact same techniques we already had shown were what you had to do to clone a dog," says Texas A&M's Mark Westhusin. "It was just the fact that you had to have lots and lots of dogs, because you have to use eggs that you collect surgically and you can't do in vitro maturation like you can with cats. They live in a country and a society where costs and animal welfare issues allow them to have hundreds or thousands of dogs on site. They can put more embryos in. It's mass production, by brute force. By brute force you can clone lots of different things if you figure out the logistics of how to get it done."

Indeed, visitors to Hwang's research facilities at the time—both those labs where dog-cloning research was under way and those devoted to learning the mysteries of human cells—describe them as resembling factory assembly lines. In the canine lab, researchers scurried about in blue scrubs and hairnets, some monitoring the ovulation cycles of dozens of dogs, some harvesting eggs from anesthetized dogs lying belly-up on gurneys, some operating video-game-like joysticks, with their eyes seemingly glued to microscopes. In the human lab, even more were at work, including Byeong Chun Lee, who directed the day-to-day dog cloning research but worked as well on the project to develop, through cloning, human stem-cell lines.

"I assume the reason that we succeeded is that we were more concentrated on the work," says Goo Jang, a member of the Snuppy research team and an assistant professor at SNU's College of Veterinary Medicine. "Dogs have a unique reproductive physiology. For example, the ovulation time in dogs. Determining when to get the oocyte was difficult. Once we knew what stage oocyte is best for nuclear transport, we could monitor the dogs and determine when we had to do the surgeries. Sometimes we did surgery at midnight to collect it. Every day at every time we were ready. Some of the North American scientists, they take holidays. But for collecting good quality oocytes, we had to follow the dog's reproductive cycle. That would set our schedule."

By improving and optimizing the procedures established at Texas A&M, and with access to far more "egg dogs"—private dog-farm owners would call the veterinary school when their dogs went into estrus—scientists at the veterinary school were able, through a vast amount of trial and error, to make Snuppy, a dog that has spent his five years of life since then living in a laboratory.

Dr. Hwang was having little problem finding donors for his human project. For his 2004 study, sixteen women donated 242 eggs in support of his effort to clone an embryo and create stem-cell lines. He would go on to claim, in a paper his team submitted to the journal *Science*, that he succeeded in creating eleven such lines.

American scientists—whose own embryo-cloning research had been stymied by a ban on federal funding ordered by President Bush—could only look on in awe, worry, or, in some cases, suspicion. With his own country's backing, though, Hwang was named director of the World Stem Cell Hub, a new stem-cell bank for humans. Cures for incurable diseases were just around the corner, he said in speeches and appearances, and would be as simple as injecting the crippled and diseased with their own, newly grown stem cells—a far more palatable solution, to Korean society, than organ transplants, which many shun in the belief that their body parts are the property of their ancestors.

In light of the promise of his human research, Hwang developed a rabid following. An "I Love Hwang Woo-suk" website popped up on the Internet. Hwang joked that he was Korea's answer to Elvis, and one fawning local media report compared Hwang's hands to those of God. Snuppy's image—dogs being far more photogenic than stem-cell lines—became Hwang's unofficial logo.

The dog Snuppy was cloned from, meanwhile—except for occasional reunions with his clone—stayed mostly out of the spotlight.

Tai was born in Michigan, and had been sold by a breeder to Cheol Yong Hwang, an SNU student, who took delivery of the purebred Afghan, at four months old, in 2002.

Though often viewed as the dumb blonde of the dog world—one American expert ranks the breed dead-last in terms of intelligence—an Afghan was the shy and quiet veterinary student's breed of choice. He prized Afghans for their independence and elegance. He kept Tai meticulously groomed and, once the dog was old enough, began entering him in dog shows. Although he was two inches taller than kennel-club standards designate as ideal for Afghans, Tai, as one of the few Afghans in the country, racked up dog show awards and, at two years old, was named a champion.

His owner would sometimes bring the dog to SNU, where, in 2005, he was a postdoctoral veterinary student. Dr. Lee, noting the then three-year-old dog's striking appearance, stopped him in the hall one day.

"He told me that he wanted to clone my dog. He asked me, 'What do you think about Tai being the first cloned dog?' " Cheol Yong Hwang was reluctant, but didn't voice any opposition. "It's Korean culture, maybe," he explained. "You follow the senior professor's opinions." He says his only request was that the tissue sample be taken from inside Tai's ear. The area to be biopsied must first be shaved, and he didn't want the dog's appearance even temporarily altered.

"To tell the truth, I wasn't sure it would be possible, the cloning of a dog," he said. Now an assistant professor specializing in internal medicine and dermatology at SNU's veterinary school, he freely admits he agreed to the cloning of Tai—short for Taipa Sahale, an Indian phrase meaning "wings spread above"—because he was afraid to say no to his superior.

Tai was never naturally bred. Neutered, nearing nine, and retired from the dog show circuit, he still shares an apartment near campus with Cheol Yong Hwang and another Afghan hound. His offspring, meanwhile, spends most of his hours in a crate in an SNU laboratory, and his eerie wails can be heard echoing through the hallways of the veterinary school.

Rail-thin, and exceedingly vocal, Snuppy seems eager to meet new people, but doesn't look them in the eye and quickly grows undisinterested. His caretaker—the same woman who performed the micromanipulation that placed Tai's cell in the anonymous egg—says he gets out three times a day, is well cared for, and is regularly groomed. Except for exhibitions and guest appearances, he rarely leaves the lab. While he is physically similar to his brindle-coated donor, and shares some of the same habits—like leaning against the wall when coming down stairs—the two dogs have vastly different personalities.

Tai's owner said that, other than in appearance, the two dogs are not at all alike—probably because they have been raised in such different environments.

"Tai is not vocal like Snuppy. Snuppy always wants to see people. But he is restricted from meeting people; not Tai," he says. Tai hardly barks at all—except when Snuppy is around. "Tai is so gentle a dog, but sometimes Tai barks and gets angry with Snuppy for no reason. It's a very interesting thing. He's not like that to other dogs, only to Snuppy."

The two dogs see each other several times a year, he says.

Tai, he says, doesn't seem to feel a connection to Snuppy, but Snuppy "seems to see Tai as special. Snuppy is very attuned to Tai. He always mimics Tai when he meets him." Even shortly after his birth, Snuppy seemed to prefer the company of Tai to his surrogate mother, the yellow Lab Simba, who belonged to another veterinary school student.

Cheol Yong Hwang was not involved in the cloning of Tai. Nor was he caught up in the veterinary school's fervor to clone a dog. He never asked about the progress of the experiments, nor was he ever updated, until one Sunday when Dr. Lee called him to his office and showed him a picture of the puppy clone. He was sworn to secrecy—because the announcement of Snuppy's birth was not for another three months. He didn't meet Snuppy until then.

Sitting in Lee's office in early 2009, with Lee at his desk, Cheol Yong Hwang chose his words carefully in an interview—but he didn't hesitate to speak his mind, in a way he didn't, as a student, five years earlier.

"For people who are considering cloning, my only answer is don't imagine you will get the same dog. . . . The impression is totally different, even if their appearance and temperament are similar, your impression of the cloned dog is different. So I do not recommend a clone if they want the same dog."

If asked again, Tai's owner says, he would decline a request to clone Tai, and he said he would never pursue cloning any dog he owned: "I would say no, actually. One is enough. Death is a natural thing. I would rather just keep the memory."

17.

Booger at Rest

■ ■ ■

> Little did I then expect the calamity that was in a few moments to overwhelm me, and extinguish in horror and despair all fear of ignominy or death.
>
> —*Mary Shelley,* Frankenstein

RIVERSIDE, CALIFORNIA

April 2006

At times, it seems to Bernann McKinney as though she's awash in a lonely sea of misfortune, pummeled by one monstrous wave after another. No sooner would she catch her breath and brace herself than the next would hit—calamity followed by catastrophe, tragedy in the wake of trauma.

The beginning of 2006 was one of those times. Booger got sick. The dog who had saved her life ten years earlier—then gone on to help her cope with her handicaps—put up another valiant fight, but this time he faced an opponent even more formidable than a mastiff on steroids.

It took a while to diagnose, in part because of McKinney's refusal to let Booger be anesthetized for tests, but after she took him to Washington State University for an exam, the verdict came in. It was genital cancer. Despite $14,000 worth of veterinary care, chemotherapy treatments, and prayers, Booger, who was never neutered, passed away in April, in a veterinary examination room with butterfly wallpaper, in his owner's arms.

"I looked at him and I said, 'You can't die, Booger. We're partners. You can't die.' I guess I thought he'd conquer cancer and we'd always be together," McKinney said. Booger, she said, seemed to know the fight was over. "Before he died, he looked at me with eyes so full of love, so full of wisdom. It was as if he knew something no one else knew. And his eyes said, 'I will be with you again.'"

Booger's death threw McKinney into despair. She stopped eating, she couldn't sleep, her sense of security was lost and her fears of the outside world returned. "When I had my Booger, I could always keep going. Booger helped me bloom and become human again, and hold up my head. When he died, I just lost the will to live. I just pined away after that. I felt so helpless I wanted to die to be with him. I wanted to die more than live."

She had so built her world around the dog that, without him, she saw little reason to go on, except one—the possibility that she could bring a duplicate of him back, in living form, to Riverside, California.

She had taken steps earlier to make that possible, after learning of a California company that was offering the service—or at least was saying it soon would be.

"The first step is to preserve your pet's DNA in our gene bank," the company's website said. McKinney took that step. Though her sole source of income was a disability check of less than $500 a month, she signed up, making an initial payment of $1,000. Days later, a "bio box" from the company arrived, containing test tubes and instructions for veterinarians on removing a tissue sample.

It took a while to find a veterinarian willing to go along—"a lot of them I called acted like I was a nut case," she says—but eventually she found one. She took Booger in for a biopsy. "It hurt him, he yelped," she says. "They pushed this thing like a pogo stick into his stomach and pulled a little chunk out of his abdomen."

The sample was sent, by FedEx, to Genetic Savings & Clone.

Once Booger died, knowing a piece of him was safely tucked away, and that it might be possible to generate an entirely new Booger from that piece, didn't keep her from grieving, but it did give her some hope—a reason, in her words, to "go on living." As for Booger's remains, McKinney had them kept in cold storage in nearby Moreno Valley, thinking she still might have him mounted or freeze-dried, as she did Tough Guy, whose lifeless mount remained in North Carolina. She wanted to be buried with Booger—"like the pharaohs did," she explained.

Whatever solace having Booger's genes banked brought her, it came to an abrupt end in the fall of 2006, when she—and about a thousand other Genetics Savings & Clone customers—received a letter announcing the company was going out of business.

"It was a horrible, disappointing letter, saying they were sorry but they couldn't clone my dog," McKinney says. "I was devastated. Lou Hawthorne wouldn't return my calls. I just cried and cried until I was sick."

Getting through Christmas was particularly difficult, because the holiday brought back memories of the dog. "I just loved Christmas with Booger. He unwrapped all his own presents. He'd take the bow and kind of toss it over his shoulder, then he'd chew the wrapping off. I had no kids, no husband. Booger was my Christmas."

Still, all hope wasn't lost. As Hawthorne's letter had pointed out, Booger's cells could be transferred and stored in liquid nitrogen at another company.

"I sent a letter to all GS&C customers in 2006, saying we were shutting down," Hawthorne said. "I told them that I was going to continue research with an international consortium, try to put the right partnerships together, and hopefully offer dog or cat cloning in the future. I told them that ViaGen was a first-rate organization that could safeguard their stored DNA if they wished. "The majority of clients—we had about a thousand animals gene-banked—decided to transfer their animals' DNA to ViaGen. Some, a small number, said to chuck it."

GS&C charged $1,000 to $1,500 for initial processing of tissue samples, and several hundred dollars a year for maintenance. It never took any payments to clone a dog—only to store the DNA. Customers were told that the technology was still being developed and was still a year or so away.

When GS&C shut down, customers were charged a transfer fee of "several hundred dollars," Hawthorne said, to have their animals' tissue shipped to ViaGen. McKinney went along with that, but when ViaGen sent her a new contract, she balked, concerned about its wording. She called ViaGen. She called Hawthorne. She called Mark Westhusin, the scientist at Texas A&M, though by then he was no longer connected with Genetic Savings & Clone or dog cloning.

After parting ways with GS&C, Westhusin and his team at A&M had gone on to clone the world's first deer. The owner of a hunting ranch in south Texas had sent him the scrotum of a trophy buck to see if it could be used for artificial insemination, thereby creating, it was hoped, more big-antlered bucks that could be shot by hunters on his fenced-in property. When artificial insemination failed, Westhusin, in collaboration with scientists from ViaGen, tried cloning, using cells from the tissue. It worked, and a fawn named Dewey was born in May 2003.

In the three years since then, Westhusin had moved on to other research, but that didn't stop McKinney from making contact.

"Bernann called me because she was concerned about a contract that ViaGen wanted her to sign. They sent out letters that were very straightforward and simple: We have your cells. They're here. We'll keep them as long as you want us to. But it also said if you don't pay your bill, then we have the right to remove your cells from storage and to discard them. Well, somehow she got it in her mind that she did not want to sign the contract because if she did, it would give them the right to discard her cells. . . .

"I explained to her they would only do that if she didn't pay for it. She was really just paranoid about this whole issue. She sent a copy of the contract to me and I reviewed it. I told her ViaGen's a good company. They'll take good care of them. If you want to ship them to a different company, you can do that.

"People who don't understand how this works, they imagine the cells in some vault and think if you take them out and move them, they're going to die or get lost in the mail, or thaw out. They get all these visions in their minds. I finally got her calmed down and told her to sign the contract," Westhusin said. "Then I told her if she really wanted to get her dog cloned, her best shot was to go to Korea."

That sent McKinney back to the Internet. "Booger's spirit was saying don't give up," she said. She read about Snuppy, the dog Seoul

National University had cloned the previous year, and began trying to contact Byeong Chun Lee, the scientist whose name was listed first in the scientific paper the team had published.

"It took me forever to locate him. Do you know how many Lees there are in Korea? But I finally tracked him down. Even though I have a thick southern accent and he has a Korean accent and could hardly speak English, it's like he understood me. He just said very quietly and very humbly, 'I can clone your dog.'"

"I was moved enormously by her story," Lee says. "I wanted to do my best for her dog and her. She liked to talk a lot, and she talked long, but I was motivated by her sincerity and her love for her dog."

Lee referred McKinney to RNL Bio, a stem-cell company that was in the process of working out a business arrangement with the SNU's veterinary school to clone dogs for profit. RNL Bio had been established in 2000 by Jeong Chan Ra, who, one year Lee's junior, had attended high school and college with him.

"I thought I should support Dr. Lee because his research will contribute to humankind," Ra said. "I also saw some possibility for a good business, but it was just a feeling." The company and university began collaborating in 2006, and in 2007 RNL hired a U.S. representative, Jin Han Hong, whose duties included rounding up dog-cloning customers. Ra referred McKinney to Hong, based in Rockville, Maryland, who flew to California to meet her. At that point, McKinney agreed to the $150,000 fee to produce a clone of Booger.

Since Booger's body was frozen, and traditional freezing doesn't make for good tissue samples, RNL needed to acquire Booger's original tissue sample—once stored by RNL's only competitor, Genetic Savings & Clone, and later shipped to ViaGen, to which Hawthorne also had connections. Hong expected that could pose problems.

Hong and McKinney eventually learned that Booger's cells were being stored in two separate locations—some at ViaGen, in Austin, some at another facility. The cells stored at each facility were of different types, in different stages, and RNL needed them both. McKinney got on the telephone again and began trying to track them down.

She ended up tracing the cells to Trans Ova Genetics, in Sioux City, Iowa, which wasn't able to immediately locate them. "I started thinking maybe they had discarded them, and were trying to cover up for it," says McKinney, who was convinced she was getting the runaround. Eventually, her call was put through to the laboratory, where a staff member agreed to look for them. "She called me back and had me on the speaker phone, and I could hear cows mooing in the background," McKinney recalled. "She said, 'I guess we found them.' I guess I won't really know until the puppies are born. They could have slipped us anything."

With the cells located, Dr. Lee flew to the United States to get them. Hong, who has a degree in microbiology, collected a sample from "the frozen Booger" to be used to compare the DNA of any cloned pups with the original.

They gathered back in Los Angeles, ice chests in hand, where McKinney insisted on accompanying the cells all the way to the gate, and watched—twenty-two months after Booger's death—as the Korean Airlines jet they were aboard taxied to the runway. Her recollection of the day is more poetic than that of RNL officials: "When they sailed into the sky on the big jet, a rainbow came through the clouds and my heart just soared, for I knew that my boy was going to come back to me once again," she says. "It was almost as if the hand of God just reached out and took the spirit of Booger into the sky and home to God. It was like He said, 'I'm taking Booger home now.'"

It was March 1, 2008, and McKinney's spirits were soaring again, almost as if she found a sort of closure in reopening the possibility that a copy of her dog would live on. "I was in such a dense grief until these Korean men came into my life," she says. "They didn't seem out for money like the other company. They have so much heart and human kindness in them. They really felt my grief. . . . A lot of people think people who would clone their dogs are a bunch of weirdos, but we're not. We're just people who loved our friends and miss them. . . . I can't wait until Booger 2 is born. I'm having to sell my home to pay for it, but that's OK, because I'll have my friend back. People wonder why I'm cloning him. Are they crazy? Don't they understand? Maybe he can come back to me and be my service dog again.

"In a few months, I fly to Korea to pick up my puppy."

Hawthorne, who by then had launched Genetic Savings & Clone's successor, Bio Arts International, says he had no regrets about losing McKinney as a customer. ViaGen transferred Booger's cells to RNL at their request, he said, and in an expedient manner, even though Hawthorne maintained his company, through holding patent rights, was the only one authorized to clone pets commercially. "ViaGen had no basis to decline to transfer the DNA, even though we suspected it would be used in a patent-infringing process."

He was happy to be done with McKinney's calls—sometimes threatening, always dramatic. "I've had a few conversations with her, and they're always very long and emotional. I just don't have time for them." Besides that, she had no visible means of being able to pay for a dog cloning in the first place.

"The woman," he said, "is impecunious."

18.

Dr. Hwang's Downfall

■ ■ ■

One can't expect to make an omelet without breaking eggs.

—*Maximilien Robespierre*

SEOUL, SOUTH KOREA

April 24, 2006

They came bearing gifts—flowers seeming to be the most popular choice for the clone who had, except for freedom, everything.

They came waving the South Korean flags—from tiny ones on sticks to giant banners, all with the traditional red and blue symbols for yin (duality) and yang (unity), merging, like two cells clinging to each other, in a perfectly formed circle.

They came to celebrate what was still, one year later, the only dog clone in the world, and to honor the multiplicity of things he'd come to represent.

Weighing a svelte sixty-four pounds, Snuppy—born in a lab and still living in one—posed for pictures, some while wrapped in a Korean flag. He enjoyed his favorite foods: ice cream and sausage and cake. But even as everyone joined in a hand-clapping rendition of "Happy Birthday," a funeral-like pall hung over the ceremony.

Woo Suk Hwang, the man widely viewed as the creator of Snuppy, the father of dog cloning, the pioneering researcher who, as his team was cloning the world's first dog, also became recognized as the first in the world to clone a human embryo, wasn't there.

Instead, he was buried under a pile of allegations that, in the winter of 2005 and spring of 2006, seemed to mount ever higher, leading to his hospitalization for stress and, a month before Snuppy's birthday, his firing from Seoul National University. By then, a university investigation had concluded Hwang's research team had fabricated and manipulated data in connection with its claims to have cloned and obtained stem-cell lines from human embryos, misused research funds, and violated ethical standards for procuring human eggs.

After a tearful public apology—"I was blinded by work and my drive for achievement"—Hwang, who, as a child, found solace in the cow barn, sought refuge in a Buddhist temple.

Even then, at SNU, the Snuppy show, subdued as it was, went on. And, after only a brief lull, so did dog cloning.

By the time of his first birthday party, Snuppy, though both Hwang and Byeong Chun Lee were on the verge of being indicted, had repeatedly been proven legit—first by SNU researchers, later in two more reviews of his DNA that were prompted by irregularities surfacing in the human stem-cell research.

At the tender age of one, Snuppy, gangly and not yet fully grown, was not just a thrice-verified clone but a multipurpose symbol—of

Korea, of Hwang's achievements, of national pride, and of the promise and hope of stem-cell research. He was hailed as an example of Korea's potential to become a global leader in technology, and called up in defense of Hwang and his government-backed research. If Snuppy was real, Hwang's supporters reasoned, wasn't it likely that Hwang's highly promising human embryo cloning claims, and the miracle cures Hwang said they could lead to, were as well?

Politics were kept mostly to the sidelines during Snuppy's first birthday party, with only a few handwritten signs hoisted in support of the deposed and soon-to-be-indicted Hwang. Organizers kept it low-key, and their remarks to the press were few and bland. "Snuppy has grown properly and we found no difference between Snuppy and other natural dogs," said Min Kyu Kim, to whom—with Lee also being named in the scandal—had fallen the task of being official spokesperson.

Lee received a three-month suspension in connection with the scandal, but was soon back in the lab at SNU's veterinary school. While it wasn't made public at the birthday party, he was already at work on cloning another dog by then—this time, a girlfriend for Snuppy.

The experiment would lead to the birth of a trio of genetically identical blond Afghans. Lee said the research was necessary to determine whether cloned dogs could reproduce (which Snuppy and two of the three clones would go on to prove later) and to see if such reproduction carried an inordinate risk of deformities, which have been reported in the offspring of some other naturally mated cloned species.

This time around, researchers used tissue from a female Afghan hound named Jessica, fusing her cells with egg cells of laboratory dogs. Around the time of Snuppy's birthday party, 167 reconstructed embryos were implanted into twelve surrogate mothers. Three of them gave birth. Bona, the first female cloned dog, was born June 18, 2006. Peace and Hope, born from other surrogates, followed in July.

Lee went on to clone Benji, the fifteen-year-old toy poodle he had promised a client he would clone back in 1997. In exchange for a donation by the owner to the veterinary school's lab, he undertook the cloning of her dog, using cells that had been gathered postmortem. "I felt very sorry for her after I had to remove the dog's uterus," Lee said of the poodle's owner. "So when we started thinking about making a small dog, I thought about her dog." The poodle clone, Benji Jr., was born in August, but with a joint problem inherited from her donor. Scheduled for surgery to correct it, Benji Jr. died during anesthesia. She had lived less than a year, and never left the lab. "Someday maybe we will try again," Lee said, noting they still have tissue of the dog in storage.

Before 2006 was over, Lee's team had also cloned two beagles, for laboratory purposes, and five wolves—two females in October for a zoo in Seoul, and three males in August for the zoo in Lee's hometown of Chung-ju, whose only wolf had died. The wolf cells were merged with enucleated dogs' oocytes and implanted in surrogate dogs.

As Lee expanded SNU's cloning portfolio, his former colleague Hwang, lab-less though he was, was making far more headlines, as he had been since November 2005 when cracks first started appearing in his human-embryo research. They were small questions at first, which, after being reported on a Korean investigative news program, only seemed to multiply.

It was in February 2004 that Hwang and his team had announced that they had successfully created a human embryo through somatic cell nuclear transfer, or cloning. It took 242 human eggs, but the researchers said they were able to produce a single human stem-cell line. Their paper was published in the March 12 issue of the journal *Science*.

In May 2005, the month before Snuppy's birth was announced, Hwang's team revealed an even greater achievement. Using 185 donated eggs, they had created eleven human embryonic stem-cell lines through cloning, according to the paper they published in the June 17 issue of *Science.*

Properly harnessed, Hwang said, the new cell lines could serve as the salvation for millions of people suffering from Parkinson's disease, Alzheimer's disease, heart disease, stroke, arthritis, diabetes, burns, and spinal cord damage. By cloning embryonic stem cells, and reintroducing the new cells into a patient, tissue damaged by trauma or rotted by disease could regenerate.

Hwang's research was the type that couldn't be easily carried out in the United States, where concerns about human cloning and the sanctity of embryos had led to restrictions, impenetrable even by the actor who played Superman, Christopher Reeve, who became a leading proponent of embryonic stem-cell research after a spinal injury left him paralyzed.

In Korea, the seemingly groundbreaking work led to a stream of honors for Hwang. He was appointed to head the new World Stem Cell Hub, a government-backed facility that, within a month after starting its registry, had seen 22,000 ill and disabled people sign up for stem-cell treatment. *Time* magazine named Hwang one of its "People Who Mattered" in 2004, stating he had "proved that human cloning is no longer science fiction, but a fact of life." *Scientific American* named him "research leader of the year" for 2005. Korea Air Lines granted him and his wife free flights for life. A "Dr. Hwang" postage stamp was issued by the government, depicting a man rising from a wheelchair.

Only a few months after the 2005 paper was published, an anonymous member of Hwang's research team had contacted *PD Su-cheop* (Producer's Notebook), a *60 Minutes*–type investigative news program on the Korean network MBC. In a meeting with the show's producer, he pointed to both research flaws and questionable ethics, prompting the news program to undertake an investigation.

It was while in San Francisco to receive an award from the World Technology Network in November—the same week that Snuppy appeared on the cover of *Time* as the "most amazing invention of 2005"—that Hwang got word that one of his collaborators, Gerald Schatten, had announced that he was ending his association with the Korean team due to concerns about the source of the eggs used in the human-embryo cloning and stem-cell research.

"My decision is grounded solely on concerns regarding oocyte donations in Hwang's research reported in 2004," said Schatten, a University of Pittsburgh researcher who was also a member of the Hwang-led team that produced Snuppy.

Organizers of the black-tie awards banquet in San Francisco adjusted the schedule so that Hwang could receive his honor, make his speech, and rush to catch a flight home. "We are moving ever closer to making stem-cell therapies a reality, and we should not waste another second, because there are families and friends who count every second," he said in accepting the award, which he dedicated to his eighty-nine-year-old mother. "Together, I believe we can improve the quality of life for all."

Hwang returned to a swirl of allegations in Seoul. On November 21, Sung Il Roh, one of Hwang's collaborators on the stem-cell project and the head of MizMedi Women's Hospital, called a news conference to admit he had paid women $1,400 each for "donating" eggs. The next day, *PD Su-cheop* aired a report alleging that Hwang had also received eggs from two junior members of his research team, whom he had allegedly coerced.

Two days after that, Hwang held a press conference in Seoul, denying he had forced his researchers into donating eggs, but offering to resign. The press conference aired on all of South Korea's major television networks, and it was followed by a huge outpouring of support for the scientist—and an outpouring as well of new female donors. They appeared at his lab in droves, often with flowers and encouraging notes, willing to sacrifice their eggs to ensure his research continued. In just seven days, on a website set up for the purpose, Ovadonation.or.kr, hundreds of women had signed up.

In the days after Hwang's press conference, eleven of the MBC news program's twelve sponsors had pulled their advertising, sending the program on hiatus; 760 South Korean women had registered to donate their eggs, including an entire high school class of 33 girls, to a nonprofit egg-donor foundation set up on Hwang's behalf; and the "I love Hwang Woo-suk" website went online.

The website praised the work Hwang had done, supported its validity, and, as the Korean national anthem played, pictured Hwang holding Snuppy—the two encircled, via Photoshop, by a ring of Korea's national flower, the rose of Sharon. Schatten, the American researcher some Koreans believed was discrediting Hwang in order to steal his technology, was depicted less favorably, his likeness shown being discarded in a murky swamp. "Professor Hwang! Cheer up! The people will look after you," the website said. "We have to open our eyes wide and protect Dr. Hwang from shrewd American doctors."

Terminating Hwang's research, many feared, would dash not just South Korea's standing as a global leader in biotechnology, but the hopes of millions of sick and disabled people yearning for a miracle cure.

"I want to stand up and walk again," Won Rae Kang, a popular South Korean singer and dancer, paralyzed after a motorcycle accident, said at one rally for Hwang. "Dr. Hwang is the biggest hope for us disabled people." Kang was one of two members of a techno-rap singing and dancing team called Clon.

Public support remained strong even in December, when the scandal broadened significantly. More doubts about Hwang's research surfaced on the Internet, in the form of anonymous postings on several websites, including one called BRIC, short for Biological Research Information Center.

A posting on the BRIC message board by an anonymous member first pointed out that research team photos supposedly representing different cloned cells were actually all the same cell. Days later, a critique of the DNA fingerprinting that had been used by the research team to prove the cloned stem-cells were patient-specific was posted, showing that some of the supposedly different samples were genetic matches to each other. That raised concerns that some of the data was fabricated, and that the researchers might not have produced patient-specific stem cells after all.

Hwang collaborator Sung Il Roh then came forward and told the news media that nine of the eleven stem-cell lines that the 2005 paper said had been established had actually been faked, at Hwang's request. All nine had come from the same source.

Hwang held a press conference the same day, insisting that the technology to make stem cells exists and that he had accomplished the feat, but acknowledging there were problems with some of the research data. He accused Dr. Sun Jong Kim, a former collaborator, of "switching" some of the stem-cell lines.

By then, several prominent scientists, including Ian Wilmut, who cloned Dolly the sheep ten years earlier, called on Hwang to submit his work for independent analysis, and Schatten asked *Science* to remove his name from the published paper. While Hwang's support from international scientists was eroding, South Korea's politicians came to his defense. On December 6, 2005, forty-three lawmakers formed a group to support Hwang, pledging to help him continue his experiments; and two female representatives volunteered to donate their eggs to Hwang's research team.

Hwang also had the backing of South Korea's president at the time, Moo Hyun Roh, who, despite a reputation for overregulating big business, spoke often of the need for scientific research to operate unfettered. "It is not possible or desirable to prohibit research just because there are concerns that it may lead to a direction that is deemed unethical," he said while giving Hwang an award at a ceremony in 2004. (Roh, who was president of South Korea from 2003 to 2008, committed suicide on May 23, 2009, while under investigation in a bribery scandal.)

Politicians also went after MBC, the network that ran the initial exposé, calling for an investigation into the use of hidden cameras and "coercive tactics" by reporters. After reporting in mid-December that Hwang had not created patient-specific stem-cell lines, the network received 500,000 angry e-mails. The pressure led the network to issue an apology, suspend the show's producers, and pull the show off the air.

Seoul National University launched an internal investigation on December 17, 2005, and announced its initial findings only six days later: Hwang had intentionally fabricated stem-cell research results. Nine of the stem-cell lines were fake, two were still under study. The investigative panel called Hwang's misconduct "a grave act damaging the foundation of science." Hwang's claim of having used only 185 eggs to create stem-cell lines was also contradicted by the panel, which indicated that more eggs may have been used in the research process.

On December 29, the panel announced that none of the stem-cell lines matched patients in the study, and on January 10, 2006, the panel revealed additional findings: contrary to Hwang's claim of having used 185 eggs for his team's 2005 paper, at least 273 eggs were shown to have been used. Hwang's team was supplied with 2,061 eggs in between November 28, 2002, and December 8, 2005. Despite his claims otherwise, Hwang had known about the donation of eggs by his own female researchers, and had personally escorted at least one donor to MizMedi Hospital for the egg-extraction procedure.

As for Hwang's 2004 claim to have cloned an embryo, the panel stated there was no proof of that, and it said the stem-cell line may have been generated by parthenogenesis, an asexual form of reproduction, more common in bugs and reptiles, in which an egg becomes an embryo without fertilization. They concluded that Hwang's team intentionally fabricated data in "an act of deception targeted to both the scientific community and general public."

But Snuppy, the panel confirmed, was a genuine clone.

On January 11, *Science* retracted both of Hwang's papers, and the next day Hwang held another press conference, apologizing again—but blaming other members of his research project for having falsified data and for sabotaging the stem-cell lines. Ten days later, he said two of the stem-cell lines had been maliciously switched with those of non-cloned embryos. That same day, the national post office stopped selling postage stamps commemorating Hwang's research. Hwang was suspended by SNU on February 9, 2006, along with six other faculty members who participated in the research, and the next month he was fired. The day after that, his title as "Supreme Scientist" was revoked.

On May 12, 2006, Hwang was indicted on charges of fraud, breaches of the country's bioethics laws, and embezzling about $3 million in research funds.

Prosecutors also brought charges against Lee, who was charged with defrauding the government out of $308,000 by inflating his research-related expenses.

Investigators said Hwang used bank accounts in the names of relatives and subordinates to receive about 475 million Korean won—more than $350,000—from private organizations, which he juggled between various bank accounts. Prosecutors maintained that he used the funds to buy gifts for his sponsors and for politicians who supported his work, and a car for his wife.

As soon as the indictments were announced, another rally was held in support of Hwang, with hundreds of protesters gathering in front of the prosecutor's office. Police blocked access to rooftops during the demonstration to prevent suicides, some of which had been attempted at past rallies. At Jogye Temple in Central Seoul, a group of Buddhist monks began a twenty-four-hour bowing ritual in support of Hwang.

During a prolonged series of trials that wouldn't reach a resolution for three more years, financial support continued to pour in to Hwang—based, in part, on the pride his pioneering efforts had brought the country, based more so on the promise his research was seen as holding for curing disease and extending human life. Much of it came from Buddhist supporters—enough to allow him to build and open a laboratory of his own.

In January 2007, while still barred by the government from conducting human cloning research, Hwang opened the Sooam Biotech Research Foundation in Yongin. Twenty researchers from SNU followed him, relocating to dormitories constructed nearby so that little time would be wasted in commuting to their seven-day-a-week jobs.

There his research continued, on dogs at least, as did his intense work schedule, with breaks only for meditation and court hearings. With each new trial, hordes of citizens, with near-religious zealotry, continued to show up to stand vigil. At one, in February 2007, a supporter, Hae Jun Jeong, set himself on fire in protest against what he saw as Hwang's persecution and burned to death in front of a statue of Sun Shin Yi, a sixteenth-century war hero.

Other Hwang supporters have sacrificed much to help his cause, leaving their jobs, businesses, and families behind—all because they believed in the promise of his stem-cell research. A farmer who met Hwang sold his cow farm to raise money for Hwang's research and now lives in a one-room studio in the city. Another man has pitched a tent at the entrance of SNU in a pro-Hwang vigil that has lasted for years. And a sixty-five-year-old farmer left his quiet life behind to buy a camper, emblazon it with banners, and equip it with a loudspeaker that plays the Korean national anthem as he drives around the country in support of Hwang.

"As a farmer," he explained, "when I look into Dr. Hwang's eyes, I can see he is not a person to tell a lie."

19.

Missy Accomplished

■ ■ ■

> Hell, I ain't God either. But bringing the dead back to life . . . that's about as close to playing God as you can get, ain't it?
>
> —*Stephen King,* Pet Sematary

YONGIN, SOUTH KOREA

December 5, 2007

Missy, as the plan was conceived ten years earlier, was supposed to be the world's first cloned dog. Instead, at least a dozen cloned pups preceded her duplication.

Her cloning was supposed to be an American achievement—a testament to our scientific know-how and fearless pioneering spirit, not to mention our knack for coming up with new schemes to pull in profits. Instead, Missy's clone, though manufactured with American parts, was a Korean import.

Dead since the summer of 2002, Missy was supposed to be a springboard to a multimillion-dollar American business. But after

Genetic Savings & Clone shut down in 2006, Lou Hawthorne—even while holding rights to the patents on the technology used to create Dolly the sheep—watched as another company, Korean-based RNL Bio, which had teamed up with Seoul National University, began taking steps to market dog cloning.

By December 2007, though, Hawthorne was back in the picture after striking up an alliance with Woo Suk Hwang, the deposed scientist who, after his firing from Seoul National University, had gone on to establish his own laboratory, the Sooam Biotech Research Foundation. Five months earlier, Hawthorne had contracted with Hwang to clone his mother's dog.

Missy was back in the picture, too—or at least her spitting image.

Hawthorne was all smiles, and more than a little repetitious—we tend to get that way when we talk to dogs—when he met Missy's clone at Hwang's laboratory outside Seoul, a scene he arranged to have videotaped.

"Hello there. You're so cute. You're so beautiful. Look at you. You're so beautiful. You look so much like somebody I know," Hawthorne said as he sat in a black vinyl chair at Sooam Biotech, the puppy licking his face and a beaming Dr. Hwang standing alongside. "Her fur is just as soft as Missy's was; it's just the same. It's amazing. It is amazing. . . . I've been in a thought mode. I've been thinking about this. The experience of being on the receiving end, it's very powerful, a very powerful experience. I'm happy to meet you too. I'm very happy to meet you. . . . Is she a fast runner?"

"Yes, yes," replied Hwang. "Maybe faster than you."

The successful cloning of Missy at Sooam Biotech "resurrected" more than just a woman's dog. It helped bring Hawthorne's Genetic Savings & Clone back to life, with a new and more serious name—BioArts International—and it put the company in a position to, with Hwang's help, do more than just bank the DNA of pets. For Hwang, meanwhile, it was a chance to start rebuilding his reputation, tarnished by criminal allegations that he had falsified his human embryo–cloning research.

While awaiting the outcome of his court case, and government permission to return to his human research, Hwang had cloned several dogs at Sooam Biotech by the time Hawthorne met him in August 2007. The meeting was arranged by Taeyoung Shin, a former student of Hwang's who went on to work at Texas A&M on the Missyplicity Project, then joined the staff of Genetic Savings & Clone. Shin went on to become a vice president at BioArts.

Cloning Missy was the first step in what would become a business partnership between Hawthorne's new company and Hwang's lab. Once competitors, at least when it came to being first to clone a dog, the two fellow Buddhists worked out an agreement in which Hwang's private lab in Korea would clone the dogs for BioArts' new program, "Best Friends Again." Hwang would handle the cells; Hawthorne would handle the sales.

Despite the more than $20 million John Sperling had invested in dog cloning by then—for research at Texas A&M and starting up Genetic Savings & Clone—BioArts, through its "Best Friends Again" program, would not be the first company to offer commercial dog cloning. RNL Bio and Seoul National University had begun collaborating in 2006, not long after Snuppy's birth, and in 2007, RNL hired a U.S. representative to begin lining up customers—the first of which was Bernann McKinney, a former customer of Genetic Savings & Clone.

Hawthorne, though not upset with losing that particular client to RNL, argued that the Korean company had no right to be cloning for profit—that it was a violation of the rights, legally held by BioArts, to the technology used to clone Dolly. RNL, meanwhile, said patents obtained by Seoul National University in connection with the cloning of Snuppy—and subsequently leased to RNL—granted them the exclusive worldwide right to clone dogs commercially.

BioArts' patent rights were held through Start Licensing, a company formed when the company that owned the technology used to create Dolly, Geron Corporation of Menlo Park, California, struck a deal with Exeter Life Sciences Inc., a Phoenix company owned by John Sperling. Exeter paid Geron $4 million up front, plus undisclosed future payments, for a 50.1 percent stake in Start Licensing, which includes all the cloning-related patents Geron inherited when it merged in 1999 with the Scottish research company that cloned Dolly.

BioArts patents couldn't restrict cloning in the name of research, but once SNU teamed up with RNL to commercially produce dog clones, they were treading on what BioArts viewed as its turf. While SNU was granted patents in connection with the improvements they made to the technology in producing Snuppy, Hawthorne says, those were meaningless in light of the broader patents his company held.

"They have, at best, a small improvement to the foundational cloning technology," Hawthorne said of SNU, "and an improvement to foundational technology doesn't give you the underlying rights to use that technology. Our intellectual property coverage is very substantial. Once the legal actions start, RNL will realize they don't have what they think they have. They either know that they don't have squat and are misrepresenting their position, or they just don't understand patent law."

After six months of exchanging threats, legal action would commence in June 2008, when Start Licensing announced it was taking steps to sue RNL Bio. RNL Bio, meanwhile, once Hawthorne announced plans that month for an online auction of dog clonings, issued a press release saying it planned to sue BioArts and Sooam, claiming Hwang was not authorized to use technology stemming from the cloning of Snuppy.

"RNL confirms that the only firm authorized to clone dogs [worldwide] is RNL Bio and any entity that wants to access such technology should license the rights from them. BioArts' patented technology is related to sheep cloning and has never been able to clone a dog for several years and tens of millions of dollars. RNL also warns Mr. Hawthorne to retract the misleading statements that he has released to media outlets to distort RNL's legal rights."

Hawthorne—who had referred to RNL as "black market" cloners—scoffed at that notion, calling it a case of "the tail wagging the dog." He warned, "If they try to fulfill any actual commercial orders in countries where we have protection, of course, Start Licensing is going to interdict."

Hawthorne has also questioned the relationship between SNU and RNL, which he said was different from the relationship GS&C had with Texas A&M, because A&M's dog-cloning work was never government-funded. "What's going on here is a wealth transfer from public to private. . . . That will get you fired in the West, but it seems to be flying below the radar over there," he said. Some of the SNU vet school scientists hold shares in RNL stock.

He questioned RNL's business acumen as well—specifically, remarks from company officials about how the price of cloning will go down. "What kind of business does that? Think about what effect that has. Everyone will wait. It guts your short-term business," Hawthorne said. "Let's say you want to sell now for $100,000, but you announce that in the near future you'll drop the price to $30,000—the only reason to do that is to assure a future. I think the whole thing is a stock play."

Seoul National University professor Byeong Chun Lee says the relationship between the company and veterinary school is legal, ethical, and no different from other university-corporate partnerships that exist around the world. "Under the contract, RNL gives us funds to support the clinic, so SNU does some of the projects when RNL asks them to," Lee said. Eventually, he said, RNL will have its own staff and facility, at which point cloning dogs for commercial purposes will no longer take place at the university. He has no plans to join RNL full-time. "I will stay here," he said. "My purpose is developing new techniques. . . . The function of the university is developing the technology, not to be a factory."

Lee's ex-colleague Hwang didn't consider his Sooam Institute a cloning factory, either, yet he would soon surpass SNU in the number of dog clones produced, among them Missy.

How many trials and errors and eggs and surrogates it took to produce Missy's clone isn't known. With the work being conducted at a private institute in Korea, that data remained secret—as it has at SNU with most of the clones created since Snuppy.

Six months after the meeting between Hwang and Hawthorne, the first clone of Missy was born, on December 5, 2007. In addition to being the first cloned dog to arrive in the United States, she was the first puppy clone intended to be a pet—even though part of a pending business arrangement. Hawthorne named her Mira, after Mir, a fictional, powerful but benevolent Korean dragon. He added an *a* to the end to make the name feminine and to distinguish her from the ill-fated Russian space station.

Mira settled in with Hawthorne and his nine-year-old son, Skye, who would chart differences in the behavior of Missy the original and Mira the copy, for a school science project he titled "Cloning Grandma's Dog." Skye, though he was only two when Missy died, reported that both the original and the clone were fond of broccoli. They also both liked snuggling and long walks, but, as he acknowledged, most dogs do. Missy, he noted, disliked camera flashes while Mira didn't seem bothered by them. Nevertheless, he concluded, the dogs were "77 percent similar" in terms of their behavior.

Two more Missy clones would be born in February 2008, and Hawthorne would post a video with all three of them, in front of his fireplace, on the "Best Friends Again" website. "We've seen excellent physical resemblance. We've also seen, surprisingly, a very high degree of behavioral resemblance as well. They're just as willful, just as smart, just as quirky, just as delightful as Missy herself was."

While the first three Missy clones went to members of the Hawthorne family, another two Missy clones would follow, born in Korea in June 2008. They were given by Hawthorne as gifts to friends, one of whom, at Hawthorne's request, began teaching the clone to respond to commands in Klingon, the language spoken in the fictional *Star Trek* universe. A linguist and member of the Klingon Language Institute, d'Armond Speers named the clone Kahless—after Kahless the Unforgettable, a Klingon emperor in *Star Trek* who helped his father return to life.

Three other clones of Missy are seldom mentioned by Hawthorne—including two that died of parvovirus as newborns while still in Korea.

Hawthorne says there were only a few slight differences in the appearance of the clones and the original Missy. Their coats weren't an exact match, but Hawthorne said the clones' colors would change and more closely resemble Missy as they matured. The presence of mitochondrial DNA—from the donor eggs—has little effect on the clone's genetic makeup but can still affect some traits. Not all of the clones had Missy's one straight ear and one floppy one, for example.

"There's some variation in the clones, but they're strikingly similar," he said. "Physically and behaviorally, I can't tell the difference between Missy and Mira. The only difference is Missy was completely focused on my mom, and Mira is completely focused on me. There was just something special about Missy. It is very much a spark that lives within Mira and her sisters. Missy was very smart. My family appreciates intelligence, but also strong will. Missy fit right in."

But MissyToo didn't.

In early 2008, Hawthorne delivered MissyToo to his mother, now in her eighties. Six years had passed since the original Missy, suffering from an inoperable growth on her esophagus, was euthanized under a magnolia tree in Joan Hawthorne's yard. She was cremated a week later.

Within days, Hawthorne issued a press release. "Missy, the beloved dog who inspired a multimillion-dollar cloning effort, was euthanized Saturday July 6, 2002, in Mill Valley at age 15." Joan would refer to the void that followed in *Fear Itself*, her third memoir: "I step outside into the fog, and sucking in the damp air causes me to cry again because Missy is not with me, is in fact dead, a month gone, and once again I know that 'with me' is just a manner of speaking. Of course she is with me in her absence, her excitement in a fog carrying me always forward, up the hill, following the creek bed, never losing sight of me, and I needn't worry because she will always lead me home."

But when MissyToo came home, Joan seemed something less than grateful for the gift, to judge from her remarks to a *New York Times* reporter in 2009. "They're not at all alike," she said. "Missy was robust and completely calm. Missy wouldn't come through my home and knock over every wine glass."

Played down as it seemed to be in the story, the flabbergasting fact was—after eleven years of research on two continents; after all the trials and errors; after all the testing, harvesting, micromanipulating, zapping, implanting, and legal wrangling; after an estimated $20 million of her friend's money was poured into creating a clone of Missy—Joan Hawthorne didn't want the dog.

20.

The Turnspit Dog

■ ■ ■

This I confess, he goes around, around,
A hundred times, and never touches ground; . . .
With eagerness he still does forward tend,
Like Sisyphus, whose journey has no end.

—*"Upon a Dog Called Fuddle" (Anonymous)*

SEOUL, SOUTH KOREA
June 2008

Although they had gone their separate ways, former colleagues Woo Suk Hwang and Byeong Chun Lee were both churning out canine clones by the summer of 2008—unfettered, and to the outrage of few.

If the two scientists, both of whom were awaiting the outcomes of criminal charges against them, weren't trying to personally best each other, the two dog-cloning companies they had affiliated themselves with—Lee with Seoul-based RNL Bio, Hwang with the American company BioArts—clearly were.

The battle, once cloning went commercial, would play out in the marketplace, in the courtroom, and in the arena of public relations, with each company having found among their first customers a client with a heartwarming tale behind their desire to clone their pet.

For RNL, it was Bernann McKinney, whose pit bull Booger saved her life, then went on to be her service dog. Booger's cloning by Lee and his team was under way at Seoul National University. For BioArts, it was Symington. Trakr's cells—after he was announced as the winner of the company's "Golden Clone Giveaway"—would, by July, be on their way to Hwang's private mountainside lab outside Seoul.

Both, unlike the majority of early pet-cloning clients, had agreed to go public with their stories, cooperating with efforts to gain positive publicity—not just for the companies but for the brave new field of commercial dog cloning as well.

That dog cloning would go commercial is exactly what animal welfare groups feared most. It would mean more animals being used for their eggs and as surrogates, more capitalizing on the grief of pet owners. And in a world already overpopulated with dogs—where millions a year are put down in America alone—coming up with a new way to create them, factory style, seemed disingenuous, if not irresponsible.

"It's an industry that operates on a flim-flam basis, with overseas laboratories, controversial researchers, and opportunistic entrepreneurs playing to people's natural affection for companion animals by pandering to a notion that, somehow, they can be with those pets again," said Wayne Pacelle, president and CEO of the Humane Society of the United States.

"Every animal is special in some way, and there's not a kindhearted person who doesn't grieve for the loss of a beloved companion animal, and perhaps even entertain the idea of seeing that animal again. But the reality is we won't be seeing the animal again in our lifetimes. . . . An animal is more than a reflect of his or her genetic code."

Animal rights groups in the United States turned their attention to cloning, in varying degrees, in the early 2000s, with the American Anti-Vivisection Society at the forefront. Between the number of animals it takes to clone a dog and the number of dogs already in need of homes, the AAVS saw cloning as a harmful, inefficient, and somewhat predatory pursuit—one that took advantage of pet owners when they were at their most vulnerable. Public sentiment seemed to support the view. Polls have shown that more than 60 percent believe animal cloning is morally wrong, and around 80 percent are opposed to pet cloning.

But to some extent, the issue was lost in the shuffle of what were seen as bigger and more pressing causes—from hoarding to puppy mills, from Michael Vick to dog-food recalls. There were more issues to address than there were resources to address them, as has been the case throughout the history of the animal welfare movement.

One of the first causes taken up by Henry Bergh, who established the American Society for the Prevention of Cruelty to Animals (ASPCA) in the 1860s, was on behalf of dogs that, generations before cloning, were caught up in a different kind of endless cycle.

For more than three centuries—until the late 1800s—many well-heeled humans owed their evenly cooked meat, and more, to the turnspit dog. As with today's pit bull, the turnspit dog was more of a type than a recognized breed—a mutt of terrier or spaniel heritage, especially bred for the task, with short legs, a long body, and shoulders narrow enough to fit inside a wooden wheel, a larger version of the sort you'd see in a hamster cage. Powered by paws, the wooden wheel spun, turning a chain drive, which led to the fireplace, rotating the meat that was placed on a spit.

The devices could be found in kitchens of estates, inns, restaurants, and pubs in both Europe and the United States. Usually, the dog would be leashed to the wheel, so that if he stopped walking he would choke. Sometimes he'd be threatened or thrashed, or a hot coal would be tossed in the wheel to speed him up.

Bergh campaigned to end that practice, but the turnspit dog remains one of the more audacious examples of how, for most of the 15,000 years since their domestication, dogs have been required to earn their keep, and of how they've been biologically shaped in ways that promoted optimum job performance—whether it was retrieving hunters' game, protecting livestock, or herding. Even lap dogs were engineered, in part, to attract fleas off the dog's owner and onto themselves.

In large part, cloning dogs is just one more extension of our "dominion" over them, a continuation of the what-they-can-do-for-us cycle that has seen them engineered at the whimsy, and for the comfort, of humans. There were three rationales behind choosing the first dogs to be cloned: to fill the lonely void of wealthy and bereaved pet owners; to create dogs for law enforcement, security, and assistance purposes; and to further scientific knowledge that could cure human diseases. In the twenty-first century, we'd be cloning dogs to mend our broken hearts—both literally and figuratively.

Early studies on the efficiency of cloning animals put the survival rate of clones at between 1 and 4 percent, and reported that deaths and deformities were the norm. In addition, cloned animals have been found to be likely to suffer from respiratory distress, hypoglycemia, weakened immune systems, developmental problems, and malformed organs. Large offspring syndrome, with fetuses developing to twice or more the normal size, has also been an issue. While the Food and Drug Administration, the federal agency that comes closest to regulating cloning, says the risks have decreased as technology matures, there has been little objective evidence—especially relating to the cloning of cats and dogs—to substantiate that.

Between the lack of oversight for the privately funded research at Texas A&M and the loose regulatory procedures in Korea, the number of animals used was impossible for animal welfare groups to get a handle on. While research companies and universities receiving government funding for projects are required to submit to U.S. Department of Agriculture inspections and to report the number of cats and dogs they use, pet cloning companies are not subject to federal or state licensing, and are not required to either uphold standards of care or keep records on how many animals they use.

What statistical information animal advocates could gather usually came from papers written by scientists and published in scientific journals. Adding those up showed a pretty dismal success rate: it took 319 egg donors and 214 surrogate mothers to produce the first five cloned dogs and eleven cloned cats—sixteen animals resulting from the creation and implantation of 3,656 embryos.

"When there's no federal oversight, or no publishing of scientific data, we have no idea how many animals are being used, or how they're being used behind the scenes," said Crystal Miller-Spiegel, the author of a 2005 AAVS report, *Pet Cloning: Separating Facts from Fluff*. "And with the cloning moving to Korea, that allows them to evade any sort of rules. We don't know how many dogs were used. We don't know how many cloned embryos they implanted, how many surrogates were used, how many pregnancies developed, how many puppies died at birth."

Pet cloning may have been pushed to the sidelines by what were seen as more pressing issues in the United States, but in Korea it has been questioned even less, and that's only in part due to its being a country where dogs—far from turning the spits—actually end up on them.

It is estimated that more than 2 million dogs are killed and consumed in South Korea every year, roughly equivalent to the number of unwanted dogs euthanized in the United States. Putting a stop to consuming dog—one in three respondents had eaten dog meat and only one in ten thought it should be banned, a 2006 survey by a Seoul TV station reported—is a far bigger priority for the country's fledgling animal welfare movement.

The vast majority of the 3.5 to 5 million dogs kept as companions in South Korea are purebreds and, given the higher status Koreans bestow on purebreds, are often spared from being sold as meat. Mutts, on the other hand, are still seen by many as part of the food chain.

"People say, 'I eat just mixed breeds, I would never eat a purebred,'" said Soyoun Park, president of the Coexistence of Animal Rights on Earth (CARE), one of only a handful of animal welfare organizations in South Korea, a country that didn't pass its first animal welfare legislation until 1991. CARE is working to end the practice of eating dog through lobbying, protests, and the sort of PETA-like antics that have led Park, an actress-turned-animal-advocate, to spend a lot of time inside dog crates, shouting for better treatment of dogs.

Her organization's members protest every summer at the Moran Market, a block-long open-air bazaar outside Seoul, where dogs can be purchased in part or in whole, live or dead, cooked or raw, for as little as $100. Cages line the street for a full city block, crammed with street dogs, mixed breeds, and farm-raised mutts. One can buy cuts of already butchered dog, or, for about $150 ($100, if you haggle) pick a live dog to be butchered. When a choice is made, the dog is pulled out with a noose attached to a stick, dragged into a nearby room, and given a fatal electrical shock with what resembles a cattle prod. It is thrown into a steel vat of boiling water to soften the meat and make its fur easier to remove. From there, it is tumbled in a dryer that removes most of the fur. A torch is used to burn off any that remains, and the dog is then butchered to order while you wait. About 25 percent of South Korea's dog meat is sold through Moran Market.

Dog is still recommended by many Korean doctors for patients who are recovering from surgery, and touted as a food that can boost energy and virility. It is served as sliced meat, used as an ingredient in soup, and even in beverages. Deeply rooted in the farming culture of Korea—China and Vietnam, as well—the practice dates back to a time when dogs were just another farm animal, and people were hesitant to eat larger species such as oxen, which were valued for the loads they could carry. Long stretches of poverty and war made it only more necessary and ingrained more deeply the belief many still hold that consuming dog meat benefits one's health. Especially popular in the hot summer months, dog meat—though it was removed from many menus during the 1988 Seoul Olympics—is served in an estimated five hundred to six hundred restaurants in Seoul alone.

Most dog meat in Korea comes from mixed breeds that are similar to the native Jindo breed in appearance. They are often referred to simply as "yellow dogs." Most are raised on farms, where they spend their lives in cages or on three-foot chains. Others end up in the market through more nefarious routes.

"There are dogs picked up as strays off the streets and dogs that were being used to breed pets but have gotten old and useless," said Park. "The way you can distinguish if it's a farm dog or a homeless dog is that those dogs that are raised at the farm won't look at a human directly. They don't want eye contact. Those who are not afraid about looking a human in the eye are usually dogs that have been raised in someone's house."

The market also resells dogs that vendors have procured from shelter operators who, while contracted by the government to care for strays, sometimes end up secretly selling them, to double their profits. Dogs that have been used in university laboratories sometimes end up in the market as well, she said. "The university scientists are supposed to euthanize the dogs after the research is done and take them to the company that cremates them, but it costs too much, so they sell them to the market," she said.

While those who take part in eating dog are in the minority in South Korea, those who support using dogs in biotech research are a clear majority—a result, she suspects, of scientists tying their research on dogs to finding cures for human disease. "They might otherwise have viewed it as a waste of money," Park said, "but because they see it as connected to human medical advances and beauty and longevity, they get behind it."

While the human relationship with dogs is an ambiguous one everywhere, the contradictions are more evident in South Korea, where it's possible to see patrons carrying their small pet dogs, à la

Paris Hilton, into a restaurant where the same species is on the menu. That some dogs have it better than others can be seen in the misfortunes of the legendary Sapsaree, a once revered Korean breed.

There was a time in Korea when only the rich could afford the Sapsaree, a breed whose name translates into "dog that roots out evil spirits." Their ownership was once restricted to royalty, later to the wealthy upper strata of Korean society. Images of the breed date back as early as 3000 B.C. and have shown up in ancient tales, songs, poems, and paintings. But Sapsarees have shown up as winter wear and in soup, as well.

The beginning of what almost turned out to be the end for the Sapsaree came during Japan's colonial rule over Korea. Hundreds of thousands of the dogs were killed each year, primarily for fur to line the winter boots and uniforms of Japanese soldiers during World War II. In 1940, the Japanese colonial government issued Order No. 26, calling for the slaughter of domestic dogs in Korea. Under the order, native Korean dogs like the Jindo and Poongsan were spared, because they appeared to have bloodlines similar to Japanese dogs. The shaggy-haired Sapsaree, though, was deemed by the Japanese to be a mutt, unworthy of protection and suitable for use as both fur and meat. According to some estimates, as many as 1.5 million members of the breed were killed over a three-year period.

By the 1960s, when a professor of agriculture in Daegu took an interest in the breed, and sent two of his assistants to scour the countryside in search of them, only fifty-two purebred Sapsarees could be found. Sung-jin Ha, a professor at Kyungpook National University, took possession of thirty of them and began trying to breed them on his farm in Gyeongsan in hopes the dogs would multiply. But by 1985, only eight breedable dogs remained.

Ha's son, Ji Hong, took over his father's task when he returned from school in the United States—with a Ph.D. in microbial genetics from the University of Illinois, Urbana-Champaign. Slowly, the numbers of Sapsarees increased, to 30 dogs by 1990, to 150 by 1992.

Ha petitioned South Korea's Ministry of Culture and Tourism to recognize the Sapsaree as a national treasure. In 1992, his request was approved and the Sapsaree Preservation Association, headed by Ha, was established. There are now nearly 2,500 of the dogs registered—500 of which live in a colony near Daegu, operated by the association. The association has developed programs in which the dogs—considered keenly intelligent and soothingly docile—are used to provide therapy for the mentally ill and people with autism. It continues to push for further protections of the breed, as well as encourage a stronger pet-owning culture in Korea. At the same time, it keeps working on increasing the breed's numbers.

Given all that, when scientists from Seoul National University approached the association in 2007 with the idea of cloning a Sapsaree—at no cost to the association—it obliged, offering up the most famous living Sapsaree of all, Blue Dragon.

Blue Dragon was a prized specimen of the breed, which comes in two colors—a creamy yellow and a blackish gray that appears blue under the moonlight. He has appeared in movies, and at festivals across the country, and was a regular on a soap-opera-type TV show. Despite his celebrity, to call him a stud would be stretching it. Blue Dragon, his trainer says, never showed a great deal of interest in breeding. "He doesn't have much of a spark that way. He doesn't like to do it," said Kook Il Han, who is both Blue Dragon's trainer and head of the Sapsaree association's breeding research center. "Now that he is twelve years old—even if he'd like to do it—he couldn't do it. But even when he was young, he didn't like to do it," Han said. Attempts to breed him through artificial insemination also failed.

With the cloning offer, the pieces seemed to fall into place. Blue Dragon could be genetically duplicated, buffeting the Sapsaree's numbers and ensuring a prized bloodline would continue. Better yet, it was to be free, as far as the association was concerned, with the city of Daegu, which has adopted the breed as its unofficial mascot, offering to pay the discounted $30,000 fee. Han and Blue Dragon dropped by Seoul National University's School of Veterinary Medicine, where tissue samples were taken for cloning.

Two clones of Blue Dragon were born in August 2007, one of which refused to eat and died. The other survived and was named Uni—the name being a reflection of hopes for unification between North and South Korea. While Uni is a different color than his donor is now, Han says Sapsaree pups traditionally undergo a change in coloring as they age. "He looks just like Blue Dragon did in childhood, and their physical characteristics are the same." Other than an infrequent road trip for exhibition purposes, Uni—like Snuppy—has lived in a crate at the laboratory on the SNU campus since birth, even though he was supposed to end up with the Sapsaree association.

"It's a delicate problem," Han explained. "The actual owner is Dr. Lee (at the SNU veterinary school). If the association wants to have Uni, they need to pay some amount of money to Dr. Lee. But the association cannot afford the money. Daegu city originally agreed to pay for Uni and give him back to the association, but somehow something came up and Daegu city changed their attitude."

Even if the association could come up with the money, Han said, he doesn't expect Uni would be much like Blue Dragon, in terms of personality and behavior. He suspects Uni, having spent his formative years in a lab crate, is a very different dog than the one from which he was cloned: "His hardware and genetics might be very similar to Blue Dragon, but, now that two years have passed, he's not going to be like Blue Dragon at all."

After cloning the Sapsaree, SNU, in conjunction with RNL, undertook the cloning of Chaser, a yellow Labrador who was the Korean customs service top drug-detecting dog. That resulted, in the fall of 2007, in seven births. Born from three different surrogate mothers, all the pups were named Toppy, a combination of "tomorrow" and "puppy."

In December 2007 came the births of six transgenic beagle clones, which glowed red under ultraviolet light—though that feat wouldn't be announced by SNU's veterinary school for another sixteen months. By then, only four remained. The four dogs, all named Ruppy—a combination of "ruby" and "puppy"—look like typical beagles in the daylight, except their claws are pink and the skin under their bellies has a reddish pigmentation.

For the cloning, Lee's team took skin cells from a beagle, treated them with fluorescence, and put them into enucleated egg cells before implanting them into the womb of a mixed-breed surrogate mother. The glowing dogs show that it is possible during cloning to insert genes with a specific trait, allowing the researchers to proceed with plans to clone dogs with Alzheimer's, Parkinson's, diabetes, and other diseases.

Five months later, the university, in conjunction with RNL Bio, went on to clone Marine, a Labrador retriever from Japan famed for her ability to sniff out cancer.

The original Marine, a six-year-old dog who was unable to conceive because of a womb disease, was able to detect eleven kinds of cancer, including breast, prostate, lung, and bladder, through either urine or the breath of patients, or even, in some cases, by merely sniffing their underwear. SNU scientist Lee, when he traveled to Japan to secure a tissue sample from the donor dog, tried to hold his breath when he was around her, he said, afraid what the dog might detect on him.

Original plans called for making two clones of Marine. One would be kept for further study by SNU. The other would—in the cloning equivalent of a stud fee—go to Marine's owner, Yuji Satoh, head trainer at St. Sugar Cancer Sniffing Dog Training Center in Shirahama, Japan. But when four fetuses developed successfully, and were born in June 2008—to be named Marine R, Marine N, Marine L, and Marine S (the first three named for the company, the last in honor of the donor's owner)—RNL was alerted to the scent of profits. The two extra dogs were put up for sale, at half a million dollars each. RNL officials said they were already in the process of cloning another thirty Marines from the original dog, and were considering someday cloning the clones and making them available, for a fee, to hospitals.

RNL Bio sees three separate markets for dog cloning: bereaved pet owners who want to replace their pets, agencies that raise and train working dogs for law enforcement and other specialized uses, and medical researchers who want transgenic dogs, those that, in the process of cloning, can be infected with human diseases for the purposes of researching cures. All, in the view of Jeong Chan Ra, president and CEO of RNL Bio, are noble causes. "With each batch of cloned dogs, we are contributing to mankind," he says. "Even with cloning a companion dog, we are contributing to human life, to someone who lost his hope after his dog died."

Equally noble, in Ra's view, and another use of the technology mentioned often by proponents of animal cloning, is the preservation of endangered species. Others contend that those achievements, such as the cloning of wolves for zoos, are dubious at best.

"To clone an animal just so it can be kept in a cage, it's kind of meaningless," said Park, the CARE president. "They're not advancing medical science or environmental causes. All they're advancing is the university's fame."

Under Lee's leadership, by the end of 2008, nearly thirty-five cloned dogs had been produced at SNU. But twenty-five miles away, in Yongin, Lee's departed colleague, Woo Suk Hwang, had produced even more—upward of seventy-five—in the year after opening his own institute.

Hwang and his team produced five clones of a champion golden retriever, followed by six clones of a beagle. After that came the cloning of Missy, at the request of his future business partner, BioArts CEO Lou Hawthorne. In June 2008, Hwang's team announced they had cloned a Tibetan mastiff, a rare breed much sought after in China—at least by those who can afford them. Sooam initially said the dogs were cloned at the request of the prestigious Chinese Academy of Sciences. The cloning resulted in seventeen puppies—all cloned from a single Tibetan mastiff, a breed whose most perfect specimens can sell for upward of $400,000. A Korean TV news program would later reveal that the Chinese Academy of Sciences had made no official request, and Sooam officials now admit that the idea came informally from friends of Hwang who worked there. After the mastiffs, Hwang moved on to clone his own fluorescent beagles—nineteen of them.

Unlike Lee, who had produced red fluorescent beagle clones, Hwang opted for green.

21.

Booger, Times Five

■ ■ ■

My little old dog:
a heart-beat
at my feet.

—*Edith Wharton*

SEOUL, SOUTH KOREA
August 5, 2008

Flight 893 touched down smoothly on a runway left glistening by rain. Bernann McKinney sailed through customs and found an RNL Bio representative holding a sign with her name on it in the spotlessly clean terminal of Seoul-Incheon International Airport.

Exuding more politeness than McKinney was accustomed to, he bowed and loaded her baggage into a limousine, which whisked her off to one of Seoul's more exclusive hotels. Settled in her room, she unpacked, ordered room service, and coaxed the wrinkles out of tomorrow's outfit—a silky new dress of cobalt blue.

Tomorrow—as hordes of media representatives would witness—

she was to meet the clones of her dead pit bull, Booger, or, as she was prone to calling them, "her babies."

For a fleeting moment in Bernann McKinney's potholed life, there seemed to be a red carpet rolled out before her. By agreeing to appear at the press conference, she was earning a significant discount. RNL had agreed to lop $100,000 off her $150,000 cloning bill in exchange for her cooperation in publicizing their feat, a markdown she needed more than anybody, including RNL, knew.

Up until then, she had been anonymous—an unnamed woman seeking to clone the pit bull who had helped her overcome debilitating injuries. Everyone assumed she was a wealthy eccentric. They were at least half wrong. McKinney, other than her $453-a-month disability check, had no income and no savings. She paid nothing up front.

Initially, she had hoped to maintain her privacy, a request RNL—initially—promised to honor. She was identified at first only as a California woman—a grandmother, some reports noted. In the one news report that included her name, it was misspelled—McKunney instead of McKinney—so no one had caught on to the secret in her past. When she boarded a jet to Korea to meet the dogs, she knew there was a chance all that might change.

She had a few misgivings about the deal, worked out with RNL over the phone—but not $100,000 worth. She had no way of raising that much, other than the vague hope of a book or movie contract. So she jumped at the offer, thinking there was a slim possibility that she could cooperate with news media and still keep her face mostly hidden. The media would be more interested in the dogs, she figured.

She still needed to come up with $50,000. But for now, with her room and most meals paid for by RNL on this visit—to meet the dogs she planned to come back and take home a month later—all she had to cover was the airplane ticket.

With the help of friends and family, she had managed to scrape up the ticket money. But she couldn't get the ticket without first getting a passport, and that proved troublesome—so much so that for a while it looked as if she might miss the press conference in which she, somewhat reluctantly, had the starring role.

"I am a wreck," she confided in a phone call the week of her scheduled departure. "They won't give me my passport. I went last week to get it. I paid extra for it to be expedited. They told me it would take three days. Now the passport office keeps saying it's 'locked.' What does that mean, locked? What am I going to do? I need to leave Sunday or Monday to get there in time for the press conference."

Four days before she was scheduled to leave, the passport office called and told her that, because of some unspecified problems, they would need the birth and death certificates of her parents and siblings, as well as her school records.

"There shouldn't be anything in my past that would be so awful I can't get a passport," she said. "Ever since 9/11 they think everybody's a terrorist. Sure I pleaded guilty to using somebody else's passport. I had a little violation back in 1978. But that's a long story, and it was a long time ago."

McKinney guessed that the delay was related to a thirty-year-old incident—one where circumstances required her to fly under the name, and with the passport, of a dead woman. "This is going to stop me from going," she said. "I'm really frantic. I've got to be there. I've got to get in there and see my babies." She called her congressman, whose office was able to clear up the matter.

Once she had her passport, McKinney's mood switched from

desperation to glee, buffeted also by some new information from RNL that she found difficult not to share. "I have something incredible to tell you, but I'm not supposed to tell you," she said in a phone call.

RNL had issued a press release weeks earlier, announcing that three clones of Booger would soon be born. What McKinney had learned days before her trip, and promised not to tell, was that there were five. Two more had been born to another surrogate dog. The first three were originally scheduled to be delivered, by cesarean, on July 28. Because of an out-of-town lecture engagement, SNU's Dr. Lee delayed their delivery to July 30. By then, a second surrogate dog in which Booger's cells had been implanted was ready to give birth as well.

McKinney wanted all five of them, and that was fine with RNL.

For the company, it was one more heartwarming twist in a story that already seemed a public relations dream. Not only were they pulling off the first "commercial" cloning of a dog—the first paid cloning on behalf of a pet owner not affiliated with the science or business of cloning—but they were providing a happy ending as well. McKinney's story seemed to have everything: a hero dog, a woman whose life it saved then helped make whole again, and a love so deep that it defied all the normal boundaries of nature, death included.

Officials of the fledgling Korean biotech company figured her story was certain to garner attention, provoke interest in other pet owners in cloning, and—not that it's needed when you have puppies on camera—show the pure and heartwarming side of the global marketing of cloned pets.

Just as with the cloning of pet cats, there were those speaking out against cloning dogs—animal welfare organizations, mostly. Others, including online naysayers, ridiculed it anonymously, seeing pet cloning as another example of bizarre behavior among the moneyed, as a pursuit both selfish and frivolous, as—to use the favored term—"playing God."

McKinney kept a close watch on the Internet chatter, and in some cases answered back. "Why do they fixate on the money?" said McKinney, who, in fact, wasn't a grandmother, and who never bore children. "Don't they have any scientific curiosity at all? Maybe they've just never loved a dog like I loved Booger."

But in Seoul on August 6—her birthday, no less—she forced those and any other negative thoughts out of her mind. It was her day. She was the star, albeit one who says she was still holding out hopes of telling the story of her dog while remaining in the shadows—in a new dress she thought brought out the blue in her eyes.

More than a hundred representatives of news organizations were there to capture the moment, to chronicle what on its surface appeared another wondrous development in the fast-paced world of biotechnology, and what, in reality, was a love story whose ending two years earlier was being rewritten.

At first, McKinney managed to stay on the sidelines, but when she met the dogs, the rapture of the moment made her forget those concerns. What seemed to her like a thousand camera flashes went off.

"Oh look, it's got his little white chest," she cried when she picked the first one up. "Oh, thank you, God."

At the press conference, McKinney said God had sent her the original Booger, "and He knew I'd be lost without him, so He sent me some more. . . . I dream of the day someday when everyone can afford to clone their pet because losing a pet is a terrible, terrible loss."

During interviews, she revealed little in the way of personal information, although it did slip out that she was a former beauty queen. By the time the BBC took its turn, McKinney had clearly given up hope of hiding her face.

She responded to questions from reporter John Sudworth while looking directly into the camera, kissing the one dog she held and nuzzling it with her nose.

"Here are the five tiny pit bull terrier puppies," Sudworth intoned, "all of them clones. . . . And I'm very pleased to say we can talk to the owner of the original dog, a dog called Booger, Bernann McKinney. Bernann, it's a lot of money you're paying to have these puppies cloned. Why would someone pay $50,000 to clone their dog?"

McKinney: "You see, Booger was a unique dog. He had the ability to reason. Booger was my service dog. He took care of me when I had a terrible ordeal. He saved my life. I found him as a little stray dog on the side of the road and I took him home. Later on, a huge mastiff-size dog attacked me, and it amputated this hand which had to be reattached, tore my leg all to pieces and my stomach. And Booger rushed in to save my life. . . ."

Sudworth: "So, a very, very special dog. Some people would say, though, that these puppies are identical genetic copies of Booger, but they won't be Booger."

McKinney: "Of course they will. Are we our parents? Do we have their genes? I'm just like my mom, even with her faults. So I'm sure dogs are the same way. But the thing about it is when Booger died I was suffering. I was pining for him. . . . And after I met these good gentlemen in Korea, they told me, 'We can clone your dog.' It was the most beautiful words I ever heard in the English language, 'I can clone your dog,' because now I can have him back."

Sudworth: "What about those people who would say there is something uncomfortable about this process, although death can be tragic it should be final, and there's something perhaps unnatural about wanting to bring a very loved pet back?"

McKinney: "Well, it's like this. Dr. Lee told me, he said only God can give him a soul. We give him his body and you give him his love and personality by re-creating the environment of the original Booger. To me that's not unnatural."

As Sudworth began his closing spiel, McKinney could still be heard in the background: "They're just like Booger, just like Booger . . ."

RNL officials used the opportunity to showcase some of its other clones, including the four clones of Marine, the cancer-detecting dog from Japan. They predicted booming times ahead for the dog-cloning business, especially once the price dropped, as they once again promised it would. RNL chief executive Jeong Chan Ra spoke to the possibility of cloning other animals, including camels for rich people in the Middle East.

McKinney stayed up late, granting more interviews, including one with NBC's *Today* show, in which she sat with a basket of Booger clones at her side, holding one up to the camera and pronouncing them replicas of the original.

Cloning, she said, was really "not that different than in vitro fertilization—couples who want a baby and do anything to have a baby. . . . That's kind of like cloning. The DNA is taken out and Booger's DNA is inserted in a surrogate mommy dog, who then has the baby." McKinney mentioned how she hoped to write a book and movie based on Booger and his cloning, and talked about her dream of opening up a training center where dogs would learn how to work with the handicapped. She even had a name for it: "Booger's Place."

After the press conference and interviews, RNL employees threw her a birthday party, complete with cake and party hats. She was exhausted by the time she returned to her hotel room, where, not long after she collapsed on her bed, the phone rang.

Still basking in the day's glow, McKinney picked it up. The caller got right to the point. When she heard the question, the color left her face, and the joy of the day immediately drained out of her. It was the call she dreaded. Whether it was the southern drawl, the beauty queen reference, or her face—that mischievous sparkle in her eyes that still shines through thirty years later—someone had made the connection.

"Can you tell me," the caller asked in a British accent, "if Joyce McKinney and Bernann McKinney are the same person?"

22.
The Story of a Mormon Beauty Queen

■ ■ ■

How deep is your love?

—*Bee Gees, 1978*

LONDON, ENGLAND
1978

In the fall of 1977, a Southern Baptist–raised, briefly Mormon beauty queen boarded a jet for London, a copy of *The Joy of Sex* and a pair of mink-lined handcuffs in her luggage. Her intention was to reclaim her one true love.

The popular sex manual, published five years earlier, was intended, like the cuffs, to aid her in that goal—to help persuade the object of her affections, a young missionary dispatched from Utah to spread the clean-living Mormon word in England, that he and she should be together forever.

No rules, no ocean, no religion were going to stop her—not even the unnecessarily complex protective underwear she says he wore. She planned to cross whatever boundaries needed crossing and get over, under, around, or through whatever obstacles she encountered. She would stretch the truth, if necessary, or abandon it entirely. As she saw it, any means were justified when the end was love.

London police saw it differently. She was arrested on charges of kidnapping the young Mormon. She was accused in court of doing things to him that, in the missionary's position, were both untoward and unwanted. And as 1977 lapsed into 1978, she spent three months in jail. Even that didn't mark a dead end to her quest; only a detour.

She picked right up where she left off, which said a lot about Joyce Bernann McKinney, and something about the times.

As years go, 1978 was a heady one. A year before, we landed a spaceship on Mars. A woman's likeness finally graced U.S. currency with the minting of the Susan B. Anthony dollar. The year also saw the debuts of the first cellular phone system, the first computerized bulletin board, and the revolutionary birth of the first "test tube" baby, Louise Brown, whose tidy conception in England required neither mother nor father to be present.

It was a time pregnant with possibility, prone—with the righteousness of the sixties well behind and the era of political correctness still ahead—to an anything-goes goofiness. The forerunner of all video games, Space Invaders, made its debut, and fending off repetitiously advancing rows of attacking spaceships would go on to consume untold hours of adolescent lifetime. At the same time, on television and in movies, space aliens would start getting portrayed in a more positive light—from the fanciful TV show *Mork & Mindy*, which premiered that year, to *Close Encounters of the Third Kind*, the movie, released in 1977, that showed us that, rather than blasting approaching space aliens out of the sky, maybe we should hear them out first, or at least make some music with them.

While the entertainment media's warmhearted portrayals may have increased our appreciation for space aliens—fiction goes as far as fact, if not further, in shaping our collective consciousness—depictions of cloning remained mostly dark and frightening.

None was scarier than *The Boys from Brazil*, the 1978 movie in which eight dozen little Hitler clones are manufactured and custom-raised in hopes of bringing another Adolf, if not an army of them, to power. In the fictional account, based on the 1976 Ira Levin book of the same name, Joseph Mengele, Auschwitz's "angel of death," who in real life conducted experiments on twins in concentration camps, turns up in South America with some spare Hitler cells, grows them into humans, and sees to it that they are nurtured in households similar to that in which Hitler grew up.

However far-fetched the movie might have seemed—when it came to cloning animals, we were only up to the frog by then—the birth that same year of Louise Brown, who, though not cloned, was conceived in a dish, brought it a step closer to the realm of possibility, as did a controversial book published that year, whose author claimed he had documented, and even helped arrange, a human cloning.

David Rorvik had served as editor of his college newspaper at the University of Montana, gone on to Columbia University's Graduate School of Journalism, and was working as a freelance science writer. His work had appeared in *Time* and *The New York Times*. His book, presented as nonfiction, recounts how he was contacted in 1973 by an anonymous middle-aged millionaire and went on to help arrange his cloning on an anonymous tropical island.

In His Image: The Cloning of a Man made front-page news and led to a *Today* show appearance by the author. It also led to a defamation lawsuit by Oxford University scientist Derek Bromhall, whose research Rorvik cited in the book. In 1968, Bromhall had transferred nuclei from embryonic rabbit cells into enucleated egg cells and reported the development of rabbit embryos. He didn't go so far as to transfer them into the wombs of surrogates. Bromhall's lawsuit claimed that the representation of his legitimate work in the book harmed his reputation. The lawsuit led to a cash settlement and an acknowledgment from the publishers—though not the author—that the book was a hoax. The same year *In His Image* came out, Bromhall was listed as having served as a consultant for the making of the movie version of *The Boys from Brazil*.

Despite those who dismissed *In His Image*, plenty of people bought it at the time—both the book and the story. Then again, in 1978, we were buying pet rocks, too.

Pet rocks were invented by a California advertising executive who found traditional pets too messy and costly. He started marketing the concept of rocks as pets in 1975, packaging them in pet "carrying cases," and sold more than 5 million of them at $3.95 a shot in locations varying from toy stores to the upscale Neiman Marcus.

The year 1978 also saw—though it was hardly a seminal event—the introduction of a strangely addictive game called Simon, a lap-size console, about the size of an old record album, with four colored buttons. The buttons light up in a random sequence, each with its own individual musical note. The player then tries to imitate the randomly generated patterns in the correct sequence—not unlike the scene in *Close Encounters*.

It was eerily similar, as well, to those push-the-buttons-get-a-treat experiments used to assess behavior in laboratory animals—except we humans, unlike laboratory rats, needed no conditioning and no reward. We did it without the payoff. We kept buying it and playing it. Maybe it was the mental challenge of mimicking; maybe it's because aping is in our genes. Boring as it can be, we find comfort in repetition. Our fondness for the familiar plays out in our childhood games—"Simon says do this"—in watching TV shows we've seen before, in the movies we pick, the books we read, the food we eat, and the music we like.

In 1978, we were a nation teetering on the edge of disco. The Bee Gees dominated the charts, mostly with songs about never-ending love. That theme, timeless as it is, ran particularly heavy in the top pop songs of 1978: "How Deep Is Your Love," "You're the One That I Want," "If I Can't Have You," "Baby Come Back." To turn on the radio in 1978 was to get an instant musical reminder of the possibility of endless love (which, in 1981, became a song, too). And why not? Everything else seemed within grasp—maybe even "Stayin' Alive" (No. 4 in 1978)—achievable with unflagging perseverance, catchy packaging, the right spin, and the proper push.

That applied to selling rocks, making babies in petri dishes, or, in the case of a lovelorn ex–beauty queen, turning into reality the future life she envisioned with a Mormon missionary named Kirk—right down to the white picket fence and the names of the blond-haired babies they would produce the traditional way.

Joyce McKinney met Kirk Anderson at Brigham Young University in Utah. He was an undergraduate. She was pursuing, in addition to him, a doctoral degree in drama.

She had moved West in 1972, after graduating from East Tennessee State University. She enrolled at Brigham Young as a graduate student and tried her hand at several beauty pageants. Twice she competed, unsuccessfully, to become Miss Provo, but she did end up winning a statewide contest in Wyoming—a state in which she never lived. McKinney became Miss World-Wyoming in 1973, part of a third-tier beauty contest—not to be confused with either Miss America or Miss USA. Miss World started in the late 1950s and ended in 1980, although there were several attempts to revive it under new names. Its most famous winner was Lynda Carter—the year before McKinney competed—who would go on to become TV's *Wonder Woman*.

McKinney, meanwhile, a state-level beauty queen title under her belt, continued her studies at Brigham Young and landed a role in a school production of *The Glass Menagerie*. A review in *The Daily Universe*, BYU's student newspaper, said she was "appropriately brash and pathetic, although occasionally inconsistent, as Amanda Wingfield, the desperate mother."

She took an interest in, and began attending, the Mormon Church, mainly in deference to Kirk, but soon decided "it was all a crock." Her faith in Kirk, though—six years her junior—didn't waver. "I was in my twenties and I was in love," she said. "I found my handsome prince, then suddenly he vanished."

McKinney portrays their early relationship as idyllic—and some of her acquaintances say that, though they had never been intimate, they were briefly engaged. Others describe the relationship as one-sided, and one Anderson's parents sought to terminate. On the advice of his family and bishop, Anderson took an overseas church assignment. He was sent to Ewell, a borough in the county of Surrey, just outside London, for a stint as a pavement-pounding Mormon missionary.

McKinney, with help from a private detective, tracked him down, talked a mutual friend named Keith May into accompanying her, and flew to London. As for what happened next, there are multiple versions. Police, based on a complaint filed by the Mormon church, say Anderson, twenty-one at the time, was confronted by McKinney's accomplice with an imitation revolver on the steps of Ewell's Church of Latter-day Saints, chloroformed, and shoved into a car. McKinney and May then drove their alleged captive two hundred miles to a remote seventeenth-century rental cottage in Okehampton in southeast England.

McKinney says that Anderson came with her willingly, and that what occurred was not an abduction but a "rescue." By then, she was convinced that the Mormon church was a "cult" that had "brainwashed" him. There are two versions, as well, of what happened inside the cottage. Both involved sex acts, bondage, see-through negligees, and torn pajamas—enough juicy bits to satisfy an intrepid tabloid reporter's wildest dreams.

Anderson's account—what he told the police and later testified to in court—has him being chained to the bed with mink-lined handcuffs, force-fed Bible scripture, urged to get married, tied spread-eagle to the bedposts, and sexually assaulted three times before he was able to escape. McKinney says Anderson came with her willingly and that, once in the cottage—which McKinney had stocked with romantic music, sexual aids, and his favorite foods—they engaged in consensual kissing.

"We started to kiss. . . . When our big moment came, he got scared," she said. "He starts crying. He actually cried. The next night was a little better. We finally made love. I had one of those sexual therapy books and we used some of the exercises to try and relax us."

McKinney says that all sex that occurred was consensual, that Anderson did not object to being restrained; to the contrary, she says, he found he functioned better under those conditions. After a couple of days in the cottage, McKinney says, they decided to get married, and that Anderson would leave the church. Anderson said in court he agreed to marriage only to escape. After leaving the cottage, McKinney says, she and Anderson spent some leisurely time in a park. They fed pigeons. They drove back to Surrey and stopped for lunch, which they were eating when McKinney says she saw a newspaper headline about a kidnapped Mormon.

McKinney said Anderson planned to go back to the mission, get his money and passport, inform authorities of his plans to marry McKinney, and get formally released from his missionary duties. They were to meet later and get married that day, she insists.

But authorities interrupted. As she approached the meeting place in her car—her wedding dress in the backseat—she was stopped by police and arrested. After several days in lockup, she was loaded back into a police van for a committal hearing. Before it pulled away, she was photographed holding a handwritten sign up to the barred back window: KIRK LEFT WITH ME WILLINGLY! HE FEARS EXCOMMUNICATION FOR LEAVING HIS MISSION AND MADE UP THIS "KIDNAP-RAPE" STORY.

The initial court hearing, in Epsom Magistrates' Court, provided even more fodder for the tabloid press, which was having a field day with the story of the Mormon kidnapped by the beauty queen. McKinney was never charged with rape, as many newspapers incorrectly reported; only abduction. In court, though, Anderson—maintaining sex acts were committed on him against his will—described precisely that, down to detailing the removal of his special Mormon undergarments, not chastity devices per se but designed to remind the wearer of his virtue and make losing that virtue at least a little more difficult.

On his second day in the cottage, that virtue was lost, he testified: "I couldn't move. She grabbed the top of my pajamas and tore them from my body until I was naked. I didn't wish it to happen. I was extremely depressed and upset after being forced to have sex."

In court, McKinney professed her undying and unconditional love for Anderson, almost as if it were a viable legal defense: "I loved him so much," she said, "that I would ski naked down Mount Everest with a carnation up my nose if he asked me to."

She insisted that any and all sex that occurred was consensual, and that Anderson concocted the abduction and rape story to keep from being excommunicated. Among Brits, the case, fueled by front-page tabloid headlines, became a hot topic of conversation, most often accompanied by winks, nudges, and debates over male physiology. Can a man be raped, that way, by a woman? Can one part of his body go along when the rest resists? As McKinney's attorney pointed out, Anderson was six-foot-four, and not exactly helpless: "Methinks the Mormon doth protest too much. . . . You have seen the size of Mr. Anderson and you have seen the size of my client," he said.

McKinney was held for three months in Holloway Prison before both she and Keith May were released, pending the trial—in McKinney's case, on grounds that included her failing mental health. They were ordered to stay in town, report regularly to the police station, and follow the terms of a court-imposed curfew. That was easier said than done for McKinney, whose notoriety, upon her release, loomed even larger.

"I get out and I'm a celebrity. I get mobbed. I'm getting letters from guys who want me to kidnap and rape them; some even gave me directions to their house. I'm getting marriage proposals, flowers," she said. "You see, the British take everything literally. They are very reserved and sexually repressed. And there I was, a southerner, saying things like 'I'm as happy as a pig in a wet mud puddle.' They were just tantalized by me and my story."

As she awaited trial, McKinney found herself on the party-scene A-list. She went to one, she says, thrown by the Bee Gees, and danced with Maurice Gibb, and Keith Moon, the drummer for The Who, whom she says she had to slap when he got fresh.

Still, her priority was "getting back with the guy I love"—at least until she got word of problems back home. With her court case nearing, she heard that someone had broken into her home in the States and stolen photographs of her. Some of the less honorable newspapers, she says, were planning to publish nude photos of her, further besmirching her reputation before trial. According to McKinney, a tabloid reporter whom she had befriended tipped her off to the plan. She says she never posed nude but feared the tabloids would—jackalope style—put her head on somebody else's naked body.

With that, McKinney decided—despite the court edict—to get out of the country.

First, she procured birth certificates of two dead people—one for her and one for May. With those, she was able to obtain passports in the names of the deceased, but with photos of her and May. Then, her theater experience coming in handy, she put together disguises—a red wig for her, a handlebar mustache for May. Using sign language, which she learned while working with deaf students in school, she made arrangements to travel with members of a troupe of deaf mimes headed to a performance in Canada.

About three weeks before their court date, McKinney and May, after making their regular visit to the police station, put on their costumes and boarded a British Airways flight from Heathrow to

Shannon, then an Air Canada flight from Shannon to Toronto. Airline officials later verified for police that McKinney and May, using hand gestures and written notes, told them they were traveling with the acting troupe. The airline even agreed to waive charges for McKinney's excess baggage—seven suitcases that she says were full of newspaper clippings about her.

From Canada, McKinney and May took a bus to Cleveland, where their trail dried up. Rather than return to BYU, McKinney went back home to Minneapolis, North Carolina, to lie low, or at least as low as someone with her personality—and tabloid reporters on her trail—could.

Three months after McKinney left London, another female was capturing the front-page headlines, Louise Brown, who, though conceived in a dish—with sperm from her father and an egg harvested from her mother—would, once born on July 25, 1978, get labeled the world's first "test tube" baby. By the end of the year, debate was still raging about creating life in a laboratory as opposed to the bedroom—about babies being conceived with the aid of microscopes and pipettes, as opposed to the joy of sex—and where it all might lead.

Concerns about human cloning resurfaced, and so did McKinney—in book form. The *Daily Mirror* assembled its coverage, some new reporting, and a collection of photos into a titillating paperback, copies of which began rolling off the printing presses before 1978 was finished.

It was titled *Joyce McKinney and the Manacled Mormon.*

23.

Bringing Home the Clones

■ ■ ■

> Pity the poor animals. They bear more than their natural burden of human love.
>
> —*George Bernard Shaw*

SEOUL, SOUTH KOREA
November 4, 2008

With the five pups cloned from Booger growing up in cages on the other side of the globe, Bernann McKinney saw the situation—as she sees most that are beyond her control—as dire.

Nearly three months after returning from her fateful trip to Seoul, the dogs were long since ready to be picked up. But she didn't have the $50,000.

As August, September, and October rolled by, she was missing out on their puppyhood, and the pups were missing the nurturing, training, and socialization they needed. Although McKinney questioned whether they were growing fast enough—she thought they looked a little stunted in the photos she was sent—they were getting close to being too big to be permitted in the cabin of an aircraft.

And that was one of two points on which McKinney, once she got the money to pay for the dogs, was not willing to budge.

She would not allow the clones to travel in the jet's cargo hold on the trip home. They were too fragile, and dogs sometimes died down there, especially on long flights. They would have to ride in the cabin, she insisted—even if airline rules prohibited it.

She was equally adamant about the butterfly release.

She'd had it planned from the minute she watched Booger's refrigerated tissue fly out of Los Angeles International Airport in March on a Korean Airlines jet. It was her way of honoring the dog that helped her through a difficult time, and welcoming into the world his cloned offspring.

When she found Booger in 1996, he was skinny and abandoned, searching through trash—a pitiful scene, except for the butterflies she says were fluttering about his head, almost halo-like. Booger was coal-black, with one white spot on his chest, in the shape of butterfly. When Booger "came back," McKinney said, there had to be butterflies. Butterflies were a symbol of rebirth and renewal, of hope and happiness, fluttering reminders of both the wonder of life and just how delicate a proposition it is. And you can order them online. Several companies provide them for releases at weddings and funerals. They arrive carefully wrapped in individual envelopes of light cardboard—and remain dormant until they are exposed to sunlight and temperatures of at least sixty-two degrees.

At first, McKinney planned the butterfly release for when she went to meet the clones in August, but it was so chaotic just getting there, she put it off. On her second trip, to pick up the dogs and bring them home, the butterfly release had to happen. It was the final scene of the movie in her head.

As she pictured it, she would stand in the sunshine, perhaps a grassy field—though those were hard to find in densely populated Seoul—and release five dozen butterflies, a dozen for each clone.

McKinney placed an order and checked into the legality of transporting butterflies on an intercontinental flight. She wanted to be sure customs officials in Korea wouldn't dispose of them—or, worse yet, detain her—if they found them in her luggage, but she couldn't find a definitive answer. One way or another, though, she'd bring butterflies, and she called to let officials at RNL know that the ceremony would have to be worked into the schedule.

RNL officials made no promises; they had all but lost patience with her by then, and, after the disclosure of the 1970s scandal, had no intention of making the transfer of the dogs a media event. They wanted the little Boogers out of there, with as little splash as possible, and they wanted to be paid. They made several calls to McKinney to remind her of that.

In September, the month she originally intended to pick up the dogs—by then old enough to receive the shots they would need for the trip—McKinney was still trying to raise the money she owed. She begged, she borrowed, and she tried to sell her story—the Booger story—to anyone who would have it, all without much luck.

She considered making a video seeking donations and putting it on YouTube, and began negotiating with the Animal Planet network about a reality series. "I hope they don't find out all about that garbage in my past," she said. "I said, 'Don't Google me, you're going to get stuff that will blow you away.' But I'm not worried about Animal Planet doing anything with my sex life."

In trying to revive one piece of her past, Joyce Bernann McKinney had inadvertently resuscitated another, and for the second time—first with the "rescue" of Kirk, then with the cloning of her dead dog Booger—she found herself embarrassingly tripped up in a headlong, some would say excessive, pursuit of love. Once she got back home from Korea, all the old stories bobbed, corpselike, to the surface. She was a laughingstock once again—the Mormon stalker, the lovesick beauty queen. Initial media interest in the story of the woman and her cloned dogs faded, and she was being widely ridiculed online.

In contrast to thirty years earlier—a more forgiving era, when she was viewed as "wild," as opposed to "sad"—she was older and heavier now, as some pundits felt the need to point out. "I know I don't look like a beauty queen now. I'm fifty-eight years old, and, well, I'm chubby," she said. "Now, on top of saying I raped a Mormon, they're saying things like, look how fat she got."

McKinney had never mentioned the scandal in her past to anyone at RNL. "These wonderful people who made Booger for me . . . they're Christian people," she said. "The first thing I saw when I walked into Dr. Ra's office was a picture of Christ on the wall. . . . How could I tell them something like that?"

Jin Han Hong, the company's U.S. representative, had done some Internet searching and made the connection. When he mentioned it to company officials, they didn't see how it had anything to do with getting her dog cloned. After it came out, they insisted that—as wild a story as it was—it didn't affect them one way or another.

But it was a crushing blow for McKinney. The one thing she didn't want was for the stories to merge—for her "heartwarming" cloned-dog story to be soiled by the 1970s scandal. That could interfere with her plans to make media appearances and seek donations that, in addition to being used to start "Booger's Place," a training center for therapy dogs, could also offset her cloning bill.

But merge they did. "Joyce McKinney: From Mormon Manacling to Dog Cloning," read the headline in *The Guardian*. "Dog-Cloner Denies Being Joyce McKinney of the Mormon Sex-Slave Case," reported *The Daily Telegraph*. "A Cloned Dog, a Mormon in Mink-Lined Handcuffs and a Tantalising Mystery," trumpeted the *Daily Mail*.

The story hit the United States the day McKinney got home. On August 9, the headline of an Associated Press story out of Salt Lake City asked and all but answered the question "Bernann McKinney, Joyce McKinney . . . Is the Woman Who Cloned Her Puppies the Same One Who Held a Mormon Missionary Hostage 31 Years Ago?"

McKinney watched as offers of TV appearances were withdrawn. At first, NBC wanted her back on the *Today* show. CBS wanted her too. And ABC was considering having Diane Sawyer accompany McKinney on the trip to pick up her dogs. But all lost interest—either because of the scandal, or upon learning McKinney didn't want to address that part of the story.

"Everybody canceled," said McKinney. "It went from a good wholesome clean family story to sex tabloid filth," she complained. "This is a sweet puppy story. How can anybody take a sweet puppy story and turn it into something so perverted?"

Initially, McKinney denied she was Joyce McKinney, as she did when the first reporter called before she left Korea. "I told him 'no,' because Joyce McKinney never existed," she said. While McKinney wasn't worried about the charges in England—the courts had long since decided tracking her down would require more effort than the case was worth—there were still unresolved criminal charges against her in the United States, which she had avoided by switching to her middle name and moving to California.

In one of those cases, she was charged with an attempted burglary and contributing to the delinquency of a minor. Authorities say she hired a fifteen-year-old boy to burglarize a home in Hampton, Tennessee, in an effort to raise money to pay for medical treatment for one of her horses. According to court records, the boy told his parents, who called the police, who rigged him up with a hidden recording device the day the burglary was to take place. After her arrest, her father posted her $2,500 bond the same night. McKinney never showed up for the January 2005 preliminary hearing. The charges are still pending, but the Carter County Sheriff's Office says that unless McKinney were to turn up locally, the county would be unlikely to track her down. Another still-pending case involves alleged threats she made to a woman in a parking lot. Still other charges—long since dropped or resolved—came to light after her identity was revealed.

In 1984, she was arrested in Salt Lake City on charges of harassing Kirk Anderson, and was ordered by a judge to steer clear of him in the future. In 2004, she was charged with cruelty to animals while trying to rig up a sling for her horse's leg so she could transport it by trailer to see a veterinarian in South Carolina. The charge, based on a complaint filed by a helper she hired and later fired, was dropped after two weeks.

In 1993—three years before she got Booger—McKinney was charged with trying to break into the Washington County Animal Shelter in Tennessee to free a pit bull named Dakota, who was scheduled to be euthanized for attacking a local couple while they were jogging. On October 28, 1993, five days after the attack, McKinney, wearing a wig and sunglasses, showed up at the shelter, offering her services as a volunteer, shelter officials say. Saying she was trying to lose weight, she offered to take dogs for walks. She started filling out a release form, scratched out what she had written, and wrote the name "Tammy Shelton" in the blank. She began walking dogs the same day, but when shelter staff noticed she kept returning to the area where the pit bull was housed, despite their warnings that it was off-limits, they asked her to leave.

Several days later, McKinney tried another route, authorities say, pulling up to the shelter after hours in a pickup truck and trying to scale the barbed-wire-topped concrete-block wall. A local TV station, which had been tipped off about the plan, videotaped the attempt from a nearby hill. Based on that, she was indicted on a burglary charge. Dakota was euthanized two weeks later. McKinney agreed to take part in a pretrial diversion program, and the charge was dismissed four years later.

McKinney was no novice when it came to civil law, either, according to court records of Avery County, North Carolina. She had sued the local CVS for not letting Booger accompany her into the pharmacy. She'd sued publications that ranged from tabloids to detective magazines to the local weekly newspaper. And she'd even sued her own father. In that case, aimed at getting insurance money, she blamed her father's negligence for the bee problem she says led to the attack by Tough Guy.

Another lawsuit sought damages for what she claimed was her wrongful institutionalization in 1994, and for injuries she received when, fleeing the psychiatric facility in her hospital gown, she was tackled and brought back by authorities. She told psychiatrists she was a former Miss Wyoming, that she was "internationally famous," the victim of a Mormon conspiracy, and was being pursued by paparazzi, "gutter-diggers," "trash-mongers," and "smut-peddlers." The psychiatrist, she says, nodded a lot and didn't seem to believe her.

While the pursuit by tabloid reporters may not have been as constant as she implies, it was no delusion. After McKinney absconded from England, British journalists—from the *Daily Mail, Daily Mirror, The Times,* and the BBC—converged on Avery County in 1978, interviewing townsfolk and camping in a meadow across from her farmhouse, hoping for sightings of her. At one point, a local reporter, Bertie Cantrell, then with the Avery County *Journal*, had them all over to her house.

"They . . . were intrigued by the beauty of our mountains, our southern accents, and mountain colloquialisms," Cantrell wrote at the time. "We were just as fascinated by the representatives of the British news media as they were by the fact than an Avery County girl could attract national publicity by following her Mormon lover to England and allegedly kidnapping him."

In 1994, a British film crew arrived, hoping to get an interview and footage for an "experimental documentary." Its working title was *Bound to Please: The Joyce McKinney Story*. McKinney maintains that members of the crew, in hopes of filming her, started the false rumor that she was considering suicide, which led to her stay at the psychiatric facility. The film apparently never got finished.

In 1997, tabloid reporters returned in hopes of fashioning "twenty years later" anniversary stories. "They would follow me wherever I went," McKinney recalled. "They would do stories about how fat I was, how I went to McDonald's, and name the number of milkshakes I ate."

By then, McKinney had become reclusive and, weary of those lurking around her property, she'd taken in Tough Guy, the trained guard dog that, though intended to protect her privacy, nearly took her life. "I withdrew into myself," she says. "Anybody I trusted would betray me, so I made no friends. I went into a world where I put my faith in those I could trust—animals."

Now, another ten years later, still trusting few, she was desperately trying to come up with the money to bring home the five clones of the animal she loved most—Booger.

She turned to her landlord and companion Elliott Brown for assistance, and again asked her father for financial help, including selling or taking a loan out on some of his land. "I don't know if he'll go along with it," McKinney said, "but what about the puppies in the meantime? They're little. They need food. They need their mom."

By the end of September, McKinney, who previously had had nothing but praise for RNL, had done a complete turnaround. She blamed the company—though she'd agreed to do publicity—for revealing her identity; she questioned the care her dogs were receiving; and she voiced other concerns that she had, since her visit, kept to herself.

"When I was in that lab at SNU, I saw stuff that made me want to throw up. They had dogs in there in tiny cages with no hair on their face. I'm still having nightmares about it. I really love animals and what I saw in there just made me sick," she said. "Snuppy's in a little pen and can barely turn around in it. There was another one with albino skin. I said, 'What's wrong with that little dog?' They said, 'We're not allowed to discuss it.' "

Mainly, though, she was concerned about her own dogs. Ecstatic as she had been upon first seeing the puppies cloned from Booger, she was worried about their health from the start, she confided. "The puppies were so thin. That worried me to death. The mother dog was just skin and bones. . . . I offered to adopt the mother dog. I even went out and bought $200 worth of food for her, dry and canned. I tried to get them to feed her, and they said she didn't need any more.

"Something else I don't understand," she continued. "They claimed there were two surrogate mother dogs, and I only saw one. I asked them where the other one was and they said she went back to a farm. I told them I wanted to adopt the mother dog so that the puppies could nurse as long as possible. They claimed she didn't give enough milk so they sent her back. I find it really strange, really strange."

In her view, every day that passed since her return put her pups in further danger.

By October, RNL stopped sending her photos of the pups, and they weren't as quick to return her calls. McKinney's suspicions heightened. The dogs, which she once described as identical, weren't. "All of the puppies have spots, but only two have spots exactly the same size and shape as Booger, and they come from two different moms. One has a white streak on his nose, and Booger didn't have a white streak on his nose. . . . I thought a clone was supposed to be exact.

"There's something wrong here. I don't even know that my puppies are clones. I asked them to give me a DNA test separate from the one they did. I'm wondering if maybe the cells they got weren't any good, so they just bred a dog to a black pit bull and tried to fake me out. They may have just bred that momma dog to a dog off the street and told me they were clones."

When Jin Han Hong, RNL's U.S. representative, called her, seeking payment of the bill, she interpreted the call as a threat to the pups. "I think they're going to kill them, or sell them to someone as freak dogs, or use them for lab experiments or something." People at RNL say they made no such threats and simply asked her to pay her bill and pick up her dogs.

By November, her father back in Minneapolis, North Carolina, managed to scrape up some of the money, and RNL had agreed to a payment plan. He sold a piece of property he had planned to let his daughter live on if she ever came back, and wired RNL $30,000. RNL agreed to let McKinney pick up the dogs and pay the rest later.

McKinney immediately scheduled her flight. She went online to procure certificates proclaiming the little Boogers to be assistance dogs, in hopes that all five would be allowed to ride with her in the cabin. She bought five carriers to tote them in and five "service dog" vests, each embroidered in Korean and English with the puppies' names, four of which she had bestowed in honor of the Korean scientists and RNL officials who accomplished the cloning—Booger Lee, Booger Ra, Booger Hong, and Booger Park. The fifth would be called Booger McKinney.

She packed leashes, harnesses, and, in connection with a secret plan, several canes. On November 4, 2008, she boarded a jet in Los Angeles with sixty dormant butterflies hidden inside her clothing.

McKinney passed through customs in Seoul easily, and once again RNL officials picked her up at the airport and took her to a hotel—this time a Holiday Inn, not paid for, not nearly as nice as her earlier accommodations, but one of the few hotels in Seoul that allowed dogs.

She picked up her dogs the day after her arrival. Not a camera clicked. No banners were hung, no speeches were made. This time RNL officials, while civil, seemed rushed. They'd designated a veterinarian on staff, Sean Cho, in charge of their companion animal department, to be her liaison, and he watched patiently as McKinney, before leaving, stood in an office and held a one-woman ceremony, releasing all five dozen butterflies through a partially open window.

Back at the hotel, McKinney was briefly happy. While some of the pups didn't seem as healthy as others, and all seemed undernourished, she thought she could detect a little bit of Booger in each one. She played with them, rationed out between them what dog food she had, and promised each of them that they would never be caged again.

She called Elliott, her friend and landlord, and reported that the dogs were highly intelligent. One, she said, had already learned how to open drawers. Exhausted, she fell asleep in the middle of the phone call. The next day, having decided she and the pups deserved better, she moved to a nicer and more expensive hotel.

It permitted two dogs, but McKinney got around the rule by bringing two pups in, leaving the room, then bringing more. Employees found her two cans of dog food. One of the pups bit her finger while she was feeding them, and one of them vomited. When the hotel discovered she had five dogs—and that all five had diarrhea—she was billed for the damage and asked to leave.

After that, McKinney bounced from hotel to hotel, checking in with one or two dogs, then sneaking the rest into most of them, until her November 10 departure date arrived. Sean Cho of RNL was to accompany her to the airport and on the flight home, ensuring that veterinary care would be nearby. He and other RNL officials told McKinney that the cargo hold of the jet was oxygen-, temperature-, and air-pressure-controlled.

McKinney would hear nothing of it, and, once at the airport, she revealed her idea. With the RNL veterinarian along, she figured, she just needed to find three people who were willing to pretend they were handicapped to carry a dog aboard the cabin. She would pay, with money supplied by her father, for their travel to America and back in exchange for the favor.

But Cho declined to take part in the ruse, which led to an argument that, for him, was the last straw. He had by then already had his fill of McKinney, with her incessant demands and her refusal to follow protocol. When she asked him to travel under the guise of a handicapped person and declined to check the dogs in as cargo, he walked out, leaving her, her dogs, and her luggage at the airport. McKinney got a cab to a hotel, and planned to try her strategy again the next day.

For each of the next ten days, McKinney took a cab to the airport (about $80 each way), searched for accomplices, and pleaded with airline officials to make an exception to their rule. Because she had three disabilities, she argued at one point, she should be allowed three dogs in the cabin.

Daily, Brown and McKinney's father tried to talk her into putting the dogs in cargo or bringing one dog home and leaving the rest with RNL to be picked up later. But she wouldn't hear of it. RNL had given up trying to help her by then—especially when it came to having a staff member pretend to be handicapped. "Her idea is based on the criminal," a representative said. "The puppies are not trained, but she has jackets and name tags that say they are service dogs."

At one point, she even turned to Lou Hawthorne for help. She called the head of BioArts, RNL's competitor, for advice on how to get her dogs home, how to be sure they were really clones, and whether he might be able to help with the expense of both.

"I don't think he likes me, and I don't like him either, but in a way we're sort of in a unique club. There's this international dog war going on over the patent. I could testify in his court case. Maybe I can help him get some negative publicity on RNL Bio. Right now, RNL is a bigger enemy to me, and if he's got any kind of heart at all, he'll want to help."

Hawthorne suggested she charter a plane and told her where she could have the clones independently authenticated.

"I'm not sure these dogs are cloned," McKinney said in one phone call from Korea. "I don't think I'm going to give them another red cent. One looks like a little Chihuahua, like a little spaceman. One has little tumors all over his feet. Two of them look exactly like Booger, and even have the butterfly mark, and another looks like him in the face."

After a second weekend in Seoul, and a good week after her original departure date, McKinney finally gave up on the plan to find passengers to pose as handicapped people and decided to begin shuttling the dogs home one at a time. She would find a kennel in Seoul, another in San Francisco, and make nine more trips back and forth.

By then, the dogs were all sick, and some were eating their own feces. McKinney had maxed out her father's credit card. Between taking the dogs and her luggage to the airport every day, searching for people willing to carry them on the plane, and finding new lodgings every night, she was exhausted and out of money.

On November 20, having placed four of the dogs in a Seoul kennel, McKinney boarded the plane with one. It wasn't a smooth flight. She was scolded by flight attendants when she took the dog to the restroom, laid out absorbent pads, and allowed him to do his business. The argument grew more heated and led to a visit from the plane's captain and threats to have her arrested when the plane landed.

By the time the jet arrived in San Francisco, nerves had settled down. McKinney caught a cab, took the first dog to a kennel, got some sleep, and boarded the next day's flight back to Korea, where she picked up a second dog and repeated the process. On her third trip, she finally found two allies, both of whom offered to hold a dog on their lap. By November 26, the process was complete. She gathered all five dogs and, rather than head back to southern California, made a stop at the University of California, Davis, where she had the dogs tested to see if their DNA matched Booger's.

By November 27, McKinney was back home in Riverside with five independently verified clones, all of which, like her, were suffering from giardia, a diarrhea-causing infection—passable from dogs to humans—caused by tiny one-celled parasites that cling to the lining of the intestines.

24. Trakr, Times Five

■ ■ ■

> Twin after twin, twin after twin, they came—a nightmare. Their faces, their repeated face—for there was only one between the lot of them—puggishly stared, all nostrils and pale goggling eyes.
>
> —*Aldous Huxley,* Brave New World

LOS ANGELES
June 17, 2009

Had they been five identical humans—fashioned in a lab from the cells of the same individual, shaped in substitute wombs, and marched out in a line to be introduced to the world—the reaction, likely, would have been a collective gulp.

But these were dogs, purebred German shepherds, who look alike anyway, and though one of the five was twice the size of the rest, they were puppies—warm and fuzzy, pure and innocent, all wearing matching red-white-and-blue stars-and-stripes collars and leashes.

On a day Governor Schwarzenegger was threatening to shut down California's debt-ridden state government, and Lakers fans were gathering to celebrate an NBA championship, BioArts was presenting to the public—and to James Symington, police-officer-turned-actor—five clones of his former police dog, Trakr.

They were trotted out all in a row—a scene more Disney-cute than Huxley-scary—all but the oldest with that awkward and floppy-footed puppy gait, veering off in one direction, then the other. As cameras whirred, they were led to their new owner, who was handed the leashes.

The press conference was held to present the dogs to the winner of the "Golden Clone Giveaway"—a contest BioArts had announced about a year earlier. In addition to offering a free cloning to the dog deemed—based on an essay contest—the most cloneworthy, in May 2008 BioArts announced the world's first online, eBay-style cloning auction: "The auction item is rarer than a 1937 Mercedes-Benz Roadster, a red diamond, or an original Shakespeare manuscript. . . . Five successful bidders who bid over $100,000 will have their best friend, their beloved dog, cloned."

The contest was at least part publicity stunt, but was intended even more, Lou Hawthorne said, to counter complaints he was hearing about pet cloning being available only to the very rich. "We figured we'd hear the usual absurdities from the Humane Society of the United States and PETA about pet overpopulation," Hawthorne said. "But from pet owners we thought we'd mainly hear enthusiastic praise. Instead we heard a lot of whining and anger from those who didn't have the means to participate in the auction."

He was confused by the reaction; it was almost as if pet owners felt entitled to cloning technology. "It's not like everybody calls Ferrari and says, 'Give me a free Ferrari, it's not fair that they're just for the rich,'" he said. "But a lot of the stories were genuinely touching. We thought we could clone one dog for free and that the contest would be a great opportunity to showcase some very exceptional dogs, and make a gesture to this whole community that couldn't afford it. What we didn't expect was Trakr. That blew us away."

To animal advocates, well aware of how a good dog story can open up both the public's hearts and pocketbooks, the planned cloning of a 9/11 hero dog was something of a setback in their campaign to portray pet cloning as wasteful and unnecessary, a calculated and coldhearted business aimed at exploiting people's grief. "Our reaction when we found out who the winner was, was 'Go figure,'" said Crystal Miller-Spiegel, who authored a report on the dangers of pet cloning for the American Anti-Vivisection Society. "How much more could you tug at the heartstrings of the American public?"

A lot, the press conference showed.

It was held at a rented Beverly Hills mansion, modest by neighborhood standards, across the street from where Walt Disney used to live, backing up onto Danny DeVito's property—at least according to the $25 map of stars' homes being sold by a vendor on a street corner two blocks away. Lou Hawthorne stood behind a podium that bore replicas of U.S. and Korean flags, a BioArts banner behind him, and, on an easel to the side, a larger-than-life-size portrait of Trakr, who had passed away two months earlier, at age sixteen.

Hawthorne said Symington's contest-winning essay "spoke not just of a moment in history, but also of an amazing—truly, objectively amazing—animal. . . . His many remarkable capabilities were proven beyond all doubt in our nation's darkest hour, and we're proud to have cloned him successfully."

Symington took the microphone only briefly, his voice cracking: "I want to thank BioArts as well as Dr. Hwang and his team for what is truly one of the greatest gifts I have ever received. Today is a day honoring Trakr, who was an exceptional police search-and-rescue dog. . . . I'm honored to have been Trakr's partner, and I'm honored to have the opportunity to continue his search-and-rescue legacy with these remarkable puppies—Trustt, Solace, Valor, Prodigy, and Déjà Vu. . . . There's never an easy time to say good-bye, but seeing these puppies certainly gives me a lot of comfort here today."

As flags waved and the puppies climbed into the lap of their teary-eyed recipient, Hawthorne read a message from the clones' creator, Dr. Hwang, the fired Seoul National University scientist who had gone on to clone dogs for BioArts: "9/11 was a terrible shock for Korean people as well as Americans," Hwang wrote. "These five clones of Trakr, who saved a human life at Ground Zero, are a gift not just to Mr. Symington but to America and the world."

The not-so-subtle tie-in BioArts had pulled off between cloning and patriotism was noted by few, among them the Center for Genetics and Society, a nonprofit bioethics organization based in Oakland: "When BioArts announced the winner of its Golden Clone contest on June 30, it broke new ground in emotional manipulation," it would point out on its website blog. "The cloning ringmasters who are trying to appropriate the 9/11 disaster for a practice that abuses pets and misleads pet lovers are simultaneously ridiculous and offensive."

More than two hundred people had submitted essays in hopes of winning the "Golden Clone Giveaway." Hawthorne went through all the entries, picked the top ten, and forwarded them to a panel of judges that included Liam Lynch, owner of a GS&C cloned cat; Jessica Harrison, who cared for dogs used in the Missyplicity Project experiments; and Joanne Greene, a San Francisco writer and producer.

Trakr, all the judges agreed, was the clear winner.

"He is above and beyond your regular service dog," said Harrison, who banked the DNA of her first dog, Hunter, an Australian shepherd she got in college. "You just don't get a police dog that can apprehend a suspect and then ten minutes later be surrounded by children and be a regular therapy dog. It just doesn't happen. And with the 9/11 stuff on top of all the other accolades in his background, there was no doubt about it."

Symington, in an interview shortly after he was announced as the winner, said he didn't expect the clone to be the same dog, but he did anticipate it would have Trakr's determination and the capacity to learn the same skills. "My hopes are that it could be a version of Trakr," he said. "As I understand it, your DNA is the blueprint of your body and your brain, and then nurture comes into play. I'm going to be so grateful even if it has 70 percent of Trakr's determination and drive. If we get a genetic double, then we're going to have an incredible search-and-rescue dog at our disposal."

Little did he expect that he would be presented with five clones of his dog.

"The rules of the giveaway were that we were going to do one, but once we had collectively realized the quality of Trakr, we had some discussion about whether we should just do one or two," Hawthorne explained. "Ultimately we decided, collectively, with Dr. Hwang and the Sooam team, that the world would be a better place with more Trakrs."

Hawthorne, in interviews with media outlets after the press conference, addressed the ongoing criticism of pet cloning, casting most concerns aside.

"There are some arguments that are worth serious contemplation, like animal welfare; there are other arguments that I think are a little absurd, like cloning contributing to pet overpopulation," he said. "We didn't create the pet overpopulation problem. It's a public-health problem. The government should be dealing with it. We have no impact on it, essentially. In the past, present, and future, our company will produce fewer dogs than would be born on the average suburban block."

Cloning, he said, "is like anything else—if it's done properly, it's defensible; if it's done sloppily, a whole host of problems can arise. But we're very mindful. We do it very carefully. We've thought through the ethics. If you look at all of the cloning being done in the world, it's being done by a range of talent—from some of the smartest people in the world down to high school students. High school students are cloning cows. In those cases, it barely works, and you get all sorts of problems. There are problems in the past and even today, but if you look just at the very best teams, it's working very well."

Hawthorne said the first Trakr clone, Trustt, had been born six months earlier—on December 8, 2008. The four other clones were born in April. A sixth Trakr clone was to remain at Sooam "for testing purposes." Hawthorne said Symington intended to train all five of the dogs he received for search-and-rescue work, possibly so that they could respond to future disasters. "So if there's an earthquake in Istanbul or a lost child in Yosemite, they could fly to the crisis area and field them as a team."

At the press conference, only a few questions arose about Hwang and his past troubles.

Hawthorne said he first met Hwang in August 2007 and—despite the pending criminal case against him—made arrangements for him to clone his mother's dog, Missy. After that, the decision was made to work as a team. "Working with Dr. Hwang has been a great privilege," he said. "He's done a beautiful job at everything he said he would do."

Press coverage of the clones' arrival was mostly positive, warm, and cuddly, and in some accounts, Trakr's 9/11 feats grew even larger. Trakr was one of two dogs that gave mild alerts, the day before, at the spot where Genelle Guzman would be rescued. But WCBS-TV in Los Angeles reported: "The dog actually pulled the last remaining survivor from the rubble itself."

As had been the case ever since the effort to clone a dog began at Texas A&M, the news media did little digging. Animal welfare concerns, the ethics and morality of dog cloning, and the people behind it went unexamined as the media focused instead on cute puppies and another amazing advancement in technology.

While Joyce Bernann McKinney's sensational past was gleefully exposed—mainly because of its sensational-ness—it went unrevealed that Symington had actually banked Trakr's cells years earlier, when BioArts was Genetic Savings & Clone, and that the oft-mentioned Extraordinary Service to Humanity award Trakr received came from the Bear Search and Rescue Foundation, whose founder, Scott Shields, had been convicted and imprisoned for fraudulent 9/11 claims.

In addition to claiming that his twelve-year-old golden retriever, Bear, was the first search-and-rescue dog to arrive at the World Trade Center, and that Bear—as he told a group of Boy Scouts—"found the most live people" at Ground Zero, Shields collected $16,000 in federal rental assistance based on false claims that he was living near Ground Zero on 9/11, when he was actually residing in Connecticut. He would eventually go to prison for the latter. But claims of his dog's heroics—while bringing donations to the foundation he formed in Bear's memory in 2004—would, for a long time, go without scrutiny. Numerous glowing newspaper articles on Bear and Shields were printed in the aftermath of both 9/11 and Hurricane Katrina, and in 2003 Shields self-published a book called *Bear: Heart of a Hero*. But it, like him, was mostly myth. Though he called himself "captain," Shields had never held any military or police rank. He was never certified in search-and-rescue, and Bear, despite his bright orange rescue vest, had no search-and-rescue training, either. Shields had not, as claimed, worked the Oklahoma City bombing, nor had he responded to an earthquake in Turkey. He was a former dressmaker who'd served time for fraud.

Those tiny details, like much else about dog cloning, remained in the shadows as BioArts—to a far greater extent than its Korean counterpart, RNL—sought to choreograph and otherwise control what came out in the media. Hawthorne, for example, while assuring that his animal welfare protocols were in effect at Hwang's Sooam Institute, wouldn't grant a request for a tour. But that didn't stop Sooam officials from allowing one.

It was four months earlier, and two clones of Trakr had already been born, both of which were lodged in a kennel on the ground floor of the mountainside institute. There, inside a lab, Hwang—identifiable only by the intense eyes that peered out over the surgical mask, the name stenciled on his laboratory slippers, and the legendary deftness of his hands as he manipulated the floppy uterus of an anonymous mutt in heat—was working on cloning the fifth auction bidder's dog.

Hwang used a laser to cut a slit into the underside of an anesthetized female dog, filling the operating room with the smell of burning flesh. He reached inside and tugged out the uterus. Quickly and gingerly he manipulated it between his fingers, like a stretched-out wad of Silly Putty, locating the egg cells he was looking for. A technician pushed a large syringe into the dog's fallopian tubes to flush out the cells, which were captured in a dish and handed off to another technician. Under a microscope, she isolated the cells as Hwang tucked the dog's uterus back in place, stitched up the incision, and moved on to a second dog, already anesthetized and waiting on a gurney in the room.

As Hwang worked on the second dog, harvesting more egg cells, the egg cells from the first dog—seven of them—were taken to another room. There, a technician, at the joystick-like controls of a micromanipulator, vacuumed the nucleus out of each of them, refilling them with skin cells from the donor dog. From there, they would be taken to another room, zapped with a jolt of electricity, and given an hour or so to start multiplying.

While an interview with Hwang was not permitted, because of his pending court case, a tour was offered.

In an indoor kennel, one run contained two clones of Trakr, one seventy days old, the other about fifty days old. Next door were the three-week-old clones of a racing greyhound that had died of cancer. Then there was a Boston terrier, cloned for another client, and a yellow Labrador retriever, scheduled to be delivered soon to the son of NASCAR's cofounder. Still another clone was expected to be born in the next week.

Up two flights of stairs were another series of kennels, more than a dozen runs on Sooam's rooftop, all occupied by clones—golden retrievers, Tibetan mastiffs, and some beagles that, in addition to being cloned, had their genes implanted with fluorescent green, diabetes, and Alzheimer's. At the far end of the rooftop row of cages, another dog stood on her hind legs, pawed at the chain link, and whined. It was a husky mix alone in her cage, about a year old, with one ear floppy, one ear straight—one of the Missy clones that Hawthorne left behind.

By the time of the Trakr press conference that summer, births had been achieved in all five clonings performed for the auction bidders. Born, too, was a new subsidiary of BioArts, to be called Encore Pet Science, which, in conjunction with Hwang's lab, would offer dog cloning to the public.

"Reliable high-quality dog cloning is finally here," Hawthorne stated in a stated press release about the new subsidiary. "Those lucky enough to have a once-in-a-lifetime dog can now see lightning strike a second time." The press release promised customers "duplicates of their beloved pooches. . . . Canine clones produced to date have a high degree of both physical and behavioral resemblance to their genetic donors. Though a dog's experience and training can't be cloned, genes strongly influence intelligence and temperament." The price of $138,500 would cover independent genetic analysis to certify each pet as a clone of the original pet. Encore Pet Science, the press release said, would also offer gene banking—$1,450, for processing and the first year of storage.

"Once a pet's genes are preserved in the Encore PetBank, they will be available for use indefinitely (many decades, at least). Gene banking is the ultimate insurance against the loss of a pet's genes due to death or unforeseen circumstances. If you don't gene-bank your pet's genes today, cloning may not be possible tomorrow."

25.

The End of the Dog Fight

■ ■ ■

> Dogs typically mark their territory with urine. . . . A dog may defend the territory by growling, barking, or assuming aggressive body language.
>
> —*Microsoft Encarta Online Encyclopedia, 2006*

MILL VALLEY, CALIFORNIA
September 11, 2009

After more than a year of posturing, sniffing around each other's business, growling, snarling, baring their teeth, and raising their hackles, the two companies cloning dogs—one American, one Korean—were headed to court, where Korean judges would determine which was the rightful heir to the pet-cloning market.

Lawyers, the human method for marking one's territory, prepared their arguments for a case that seemed destined to be long and complex. It would likely result in booting from the business one of the two companies, both of which had by then invested millions of dollars in pet cloning.

BioArts, Lou Hawthorne's California-based company, claimed it was solely authorized to clone pets, by virtue of its holding rights to patents issued in connection with the cloning of Dolly the sheep. The Dolly, or Roslin, patents were the property of Start Licensing, a Texas-based company partly owned by John Sperling. Start had granted rights to the patents to BioArts.

When RNL Bio announced in Korea that it had signed a contract to clone a California woman's pit bull—its first commercial cloning—Hawthorne, who had yet to deliver a cloned dog, other than his own, sounded his first warning. He called RNL a "black market" cloning company and predicted Start Licensing would, in addition to pursuing legal action, seek to bar any attempt by RNL to send a commercially cloned dog to the United States.

The Korean company, headed by a Baptist, would continue to test Hawthorne's Buddhist-inspired patience—even after it handed Bernann McKinney five clones of her deceased pit bull, Booger. RNL Bio CEO Jeong Chan Ra maintained that his company alone was authorized to clone dogs commercially, by virtue of patents filed in connection with the creation of Snuppy. The Snuppy patents, which were based on improvements to the technology that led to the creation of the Afghan clone, were owned by Seoul National University, with rights leased to RNL. While Lou Hawthorne and the Hwang-led cloning team at Sooam, with whom Hawthorne had contracted, could clone and sell all the sheep they wanted, dogs were RNL's domain, he asserted.

RNL announced bargain rates and assured the price for dog cloning would go down in the future. It continued signing up new clients. It even, at one point—before Hwang had succeeded in cloning the Hawthornes' family dog—offered to clone Missy for him, pointing out how all of Hawthorne's efforts, from those at Texas A&M to those in collaboration with Hwang, had failed.

There was some consideration given to joining forces. Hawthorne made overtures to the Korean company and met with their officials at least once, according to sources at RNL, but—between what the Koreans perceived as his arrogance and his demands to be in control of the operation—no truce was reached, and the fight went on. Like two dominant dogs vying for a single bone, both companies seemed willing to go to whatever lengths and costs necessary to seize the prize—a monopoly on pet cloning.

First, Start Licensing, after several letters of warning, sued RNL, claiming the Korean company had no right to be using somatic cell nuclear transfer to clone dogs. RNL countered by suing Woo Suk Hwang and the Sooam Institute, saying the technology Hwang was using to clone dogs for BioArts was the exclusive property of Seoul National University. Having been fired from SNU, Hwang had no right to be using the technology for commercial purposes, RNL officials said.

It was all very confusing, most of all to potential dog-cloning clients—a small, generally well-heeled, often eccentric niche, not seen, at least by Hawthorne, as large enough to support two companies. Customers interested in getting a laboratory-made replica of their dog faced the possibility that, whichever company they chose, it might be barred by a court decision from cloning pets, possibly even before the cloning was achieved.

The situation was further complicated by the sobering fact that, on top of the patent infringement case, there were still criminal cases pending against the two chief cloners—SNU's Lee and Sooam's Hwang. One or both, before 2009 was over, could be headed to prison in connection with the human-embryo research done in 2004 and 2005, which had been deemed fraudulent.

Hwang faced the more serious charges—fraud, misusing $2.25 million in government funds, and violating bioethics laws. Prosecutors were asking for a four-year sentence for Hwang upon conviction. Lee faced lesser charges; for him, conviction would likely lead to a shorter prison sentence, if any.

Still, a prison sentence for either would be close to disastrous—not just for their careers as scholars but for the companies with which they'd become affiliated. RNL's dog cloning revolved around Lee; BioArts, under its subsidiary Encore Pet Science, depended on Hwang to get dogs cloned.

While both had trained teams working under them, which would be able to continue cloning dogs in their absence, little actual cloning seemed to be accomplished without them. In 2009, both men had seen the time they could devote to dog cloning diminish—mainly due to court appearances, both for the criminal-court case and the patent-infringement case.

Both, as well, were pursuing other goals. Coincidentally or not, Hwang and Lee had almost simultaneously applied their skills to cloning pigs. In May, Hwang's Sooam Institute announced it had created a cloned pig embryo, extracted its stem cells, and established lines of self-reproducing cells from them—basically, the same achievement he announced with humans in his later discredited 2005 paper. Not to be outdone, Lee's research team at Seoul National University announced in September that, in conjunction with Hanwha L&C Corp., a Korean company that manufactures construction materials, it had cloned a transgenic pig that could lead to treatment of diabetes in humans. The pig clone, named Ispig, was born on September 19.

It was almost as if Lee and Hwang, who have had little contact with each other outside courtrooms since 2005, were in a silent competition, one attempting to out-clone the other, with neither paying much heed to the mistakes, warnings, deaths, and deformities that came up along the way. Both still wanted to clone a Siberian tiger. Lee was interested in trying to clone a camel. Hwang, it was learned through his own court testimony in the criminal case, had used government funds to pay the Russian Mafia to provide him with tissue samples from prehistoric mammoths, extinct for millions of years. He tried three times to clone one, he testified, but the experiments failed.

The last project Hwang and Lee had worked on together, shortly after Snuppy, was the cloning of a wolf. The study they published under both their names would later be found to have errors in its data and get pulled from the journal in which it appeared, *Cloning and Stem Cells*, whose editor in chief was Ian Wilmut. But the first two wolves born through cloning, Snuwolf and Snuwolfy, born in 2005, were verified as legitimate. In August 2009, before reaching the age of four, Snuwolf died at the zoo where both clones lived.

Snuwolf was found dead in her enclosure at Seoul Grand Park, her mouth covered with blood. Initially, the death was attributed to "natural causes," even though wolves generally live ten to twelve years. A necropsy was performed, and a month later, Lee reported that the cause of death was still undetermined. Lee said hot weather had decayed Snuwolf's body. "I could not get the biopsy results and single out the bacteria that caused the death . . ." he said. "We must not jump to a conclusion."

By fall of 2009, South Korea was still the only country in the world in which dogs were being cloned. Unmonitored, virtually unrestricted, and aided by the never-ending supply of farm dogs, Hwang was filling orders for Encore Pet Science at Sooam Institute, while Lee and Seoul National University, under an arrangement with RNL, were taking new orders as well—all while waiting for the outcome of the court cases.

Before any of the legal matters were resolved, though, Lou Hawthorne, in a surprise announcement, suddenly bowed out of the dog fight. The news came out not on Valentine's Day but on September 11, 2009.

Hawthorne issued a press release and lengthy statement announcing BioArts' withdrawal from the market. He lashed out at his competitors and at his own licensing company, whose failure to wage a strong fight to uphold his patent rights played a large role, he said, in his decision.

"After a lengthy and expensive negotiation, it became apparent that Start was unwilling either to commit to defend their cloning patents against infringers or to grant to BioArts the right to do so on their behalf. Start was afraid to defend their patents against challengers in the dog cloning space because, if they lost, they might also lose the ability to control markets they actually cared about—mainly agricultural cloning. Start's strong preference was to do nothing to defend the dog cloning market against patent infringers. . . . Privately, we were appalled by what we perceived as the gutlessness of our licensor, a critical partner without whose support our success was essentially impossible."

While Start Licensing issued statements to the media and filed a lawsuit against RNL, it made no effort to intervene in McKinney receiving her clones from RNL. Its statements were "late, weak, and poorly promoted," Hawthorne said, and the licensing company's legal response was "comparably anemic."

Hawthorne reserved most of his vitriol, though, for RNL, once again referring to it as a "black market" company, and once again turning to Ferrari analogies. Hawthorne had previously equated his company to Ferrari and RNL to Volkswagen. While bowing out of dog cloning meant the Volkswagen won the race, Hawthorne expressed doubts about whether it would survive in the long term. By promising prices for dog cloning would drop in the near future—from $150,000 to $30,000—RNL effectively killed what little immediate market there might have been, killed Encore Pet Science, and possibly signed its own death warrant, Hawthorne said.

"Who in their right mind markets a luxury service by announcing that the price will soon fall by 80 percent?" he wrote. "Imagine if Ferrari used the same strategy: 'Our new sports car lists for $200,000, but will soon be available for $40,000.' Obviously, customers would all wait for the cheaper product."

He added, "It's possible that RNL's grand strategy is to take over the dog cloning market by first eliminating all profit, rather like the Air Force major in Vietnam who deemed it necessary to destroy a certain village 'in order to save it.' This is not as crazy as it sounds; from deep in the Korean psyche comes the phrase *Nuh jook go nah jook jah*, loosely translated as 'If I die, I am taking you with me.' All Koreans know this phrase, which arguably lies at the heart of the fifty-six-year standoff between North and South Korea."

Hawthorne said that the dog cloning is a small "luxury" market, incapable of surviving if it offers customers cut-rate prices. Low-priced cloning, he argued, is only possible if a company isn't following animal welfare standards.

"At BioArts, we don't believe that the Western market—the largest for dog cloning—will (or should) purchase cloning services in the absence of strong welfare protocols, which is why we've decided not to compete at such low price points." Hawthorne, who had instituted a strict animal welfare regimen at Texas A&M, insisting that homes be found for all dogs after they were used in the cloning process, claimed that he required similar procedures be placed in effect at Sooam. No adoption program was under way in February 2009, and Sooam officials said at the time that all dogs used as donors and surrogates were returned to the suppliers when their services were no longer required.

The supply of farm dogs, Hawthorne said, was ultimately what allowed South Korean scientists to beat Americans in the quest to clone a dog.

"Why were South Korean scientists the first to clone dogs? Is it because they are so much more talented than cloning scientists from other countries? . . . The answer is that cloning dogs has far less to do with scientific acumen and far more to do with the availability of dogs as ova donors and embryo recipients (surrogate mothers). At current cloning efficiencies, an average of twelve dogs are needed as donors and recipients to produce a single cloned puppy. It is only possible to master canine cloning in a country where dogs are very plentiful, as they are in South Korea. But why exactly are dogs so plentiful in Korea? . . . A small minority in South Korea—as in North Korea, Taiwan, East Timor, China, and Japan—also eat dogs occasionally. . . . South Korea has an industry that raises a certain breed of dog as food, resulting in large numbers of these dogs also being available for use in cloning."

Hawthorne's statement also included his first public reference to the deformities and abnormalities that are part and parcel of dog cloning. Those, he acknowledged, have been commonplace in cloning programs at both BioArts and RNL.

"Unfortunately, in addition to producing and delivering numerous perfectly healthy dog clones, we've also seen several strange anomalies in cloned offspring. One clone—which was supposed to be black-and-white—was born greenish-yellow where it should have been white. Others have had skeletal malformations, generally not crippling though sometimes serious and always worrisome. One clone of a male donor was actually born female (we still have no good explanation for how that happened)." He didn't specify what became of those animals.

Also of concern to Hawthorne—or at least of concern once he stepped out of the business—was the surplus of dogs cloning often produced. Because of the repeated efforts required for a single successful cloning, both companies, while trying to produce one clone, were ending up with more. "Most clients only want one or two clones at most," Hawthorne said. "What are we supposed to do with the rest?" He didn't specify what became of those, either.

Despite all the reasons Hawthorne listed, some still speculated whether other factors were at play in the decision to shut down Encore Pet Science—whether the looming possibility of a prison sentence for Hwang played a role or if John Sperling, after nearly a decade of financially supporting the American quest to clone a dog and, later, Hawthorne's business, had pulled the plug.

"Animal cloning is a race to see which group can turn cloning into a financially viable business," Sperling had written in *Rebel with a Cause*, his 1998 autobiography. "This scientific race is not quite as noble as mapping the human genome, but to the scientists involved, it is equally compelling. . . . GS&C was just the first step in a potentially gigantic business."

With GS&C's successor, BioArts, pulling out of cloning, that business plan was dead—and unlikely to be revived again. The race was lost, and his multimillion-dollar investment produced—other than a half-dozen Missy clones—little in the way of a return. For Sperling, it was a rare failure, one that it's safe to assume—he declined repeated requests for interviews—left him less than pleased.

In his closing statement, Hawthorne also bemoaned the "distraction factor"—annoying questions from the media and bloggers, most of them dealing with the wisdom of cloning dogs when there are so many homeless ones. "Blaming cloners for pet overpopulation is like blaming Ferrari for the traffic on the nation's highways," he said. "If the market were larger, this media attention would be most welcome, but when it consistently fails to result in sales, it's just a distraction."

Hawthorne concluded that his company had better and more profitable things to focus on—projects that offer "both dramatically greater ROI (return on investment) and humanitarian benefit than pet cloning. Why waste time applying our new technologies to a small and problematic industry like pet cloning when they're worth so much more in other fields?"

A week after the BioArts announcement, a court ruled in Hwang's favor in the patent infringement case filed by RNL. The ruling by the Seoul District Court agreed with Hwang's contention that the technology he was using at Sooam was substantially different—namely because he had more than doubled the voltage of the electric jolt SNU prescribed to stimulate donor and egg cell to fuse. For Snuppy, 3.5 kilovolts per centimeter had been applied; at Sooam, Hwang was using about 7.5 kilovolts. The court ruled that constituted a sufficient enough variance to be considered a new technology, and that Hwang therefore was not infringing on RNL's patents. RNL said it planned to appeal the verdict.

Looming much larger was the criminal case. As a verdict neared, Hwang's supporters rallied again, demonstrating outside the courthouse. Fifty-five elected officials signed petitions that urged leniency and called for allowing Hwang to continue his research because of its importance to South Korea's future as a biotech leader.

On October 26, 2009, the verdict came down—all 250 pages of it. The trial had included forty-three court hearings, held over three and a half years, with testimony from sixty witnesses and 780 pieces of evidence entered, including 20,000 pages of investigative documents. The final ruling took the judge two months to write.

Hwang was found guilty of illegally paying women for their eggs and of embezzling government research funds. He was not—even though the court deemed his research fraudulent—convicted of fraud.

"Despite his purposes, which may have been purely scientific, Hwang is still guilty of embezzling 830 million won (about $700,000) through forged documents and illicitly trafficking human egg-cells," the judge wrote.

However manipulated and fraudulent his research may have been, the court ruled, that isn't a crime—unless data is manipulated to gain grants and funding. The court found that two companies that donated $1.7 million to his research had done so voluntarily, and not as a result of fraudulent data presented by Hwang.

"He feels deeply sorry that this case elicited so much criticism in the scientific field and shocked the public . . ." the judge said. "His wrongdoing is not minor but does not merit the severe punishment of a prison sentence."

Hwang received a two-year sentence, suspended for three years—basically, the equivalent of three years' probation. Lee received no actual prison time, either. His penalty for pocketing government research funds was imposed in the form of a fine: 30 million won, or about $24,000.

While further appeals were still possible, the seemingly unending scandal that "devastated Korea's national psyche and brought international shame" was mostly over, the Korea *Herald* said in an editorial the day after the verdicts.

Calling the conviction "an admonition to Korean society," the editorial stressed the need for oversight of cloning research—especially when both the researchers and the public are so swept up by its promise. South Korea's scientists and citizens alike, the editorial noted, were "enthralled by an example of fast-track accomplishment, which still seems to be the foremost goal of everyone, whether in business, academia or any other fields."

Epilogue

■ ■ ■

> Men have forgotten this truth, but you must not forget it. . . . You become responsible, forever, for what you have tamed.
>
> —*Antoine de Saint-Exupéry,* The Little Prince

"Welcome to RePet, where love means no surprises."

With those words, narrated by a smooth-talking pitchman in a lavender shirt and green sports jacket, the infomercial begins:

"Your pet doesn't want to break your heart. Thanks to RePet, he doesn't have to. We can clone your four-legged loved one in just a few short hours. How can we do it? It all begins with the growing of blanks, animal drones stripped of all characteristic DNA in embryonic tanks at the RePet factory. In stage two, your pet's DNA is extracted from a lock of fur or a drop of blood and then infused on a cellular level into the blank. In the final stage, using RePet's patented cerebral recording process, all of your pet's thoughts, memories, and instincts are painlessly transplanted via the optic nerve. . . . Your cloned pet is exactly the same as he was before, right down to the

DNA, with all training and memories intact. You and your child will never know the difference. . . ."

But wait, there's more.

"Cloning is just the beginning of the story here at RePet. We also offer many genetic engineering options to make your experience with your pet even better than before. Are you allergic to your cat? We can make your RePet cat hypoallergenic. Wish your dog was smaller or larger? We can do that, too. And let's say you've just redecorated your home. We can coordinate your pet's colors and markings to match your new decorating scheme. . . ."

Minus the installed memories, the quick turnaround time, the use of drones to create a sized-to-order clone, and the color-coordination option, the concept of RePet—a fictional, mall-based pet-cloning chain in the science fiction movie *The Sixth Day*—wasn't too far removed from that behind Genetic Savings & Clone, formed ten months before the movie's release in 2000.

In the movie, set in the year 2015, Arnold Schwarzenegger's character teams up with his clone to bring down a renegade cloning operation that has secretly moved on from cloning pets—a money-losing proposition intended to soften up the public to the idea—to cloning people.

In real life, Schwarzenegger would go on to become thirty-eighth governor of California. Genetic Savings & Clone would go on to become BioArts, whose subsidiary Encore Pet Science, was the only pet-cloning company in the United States until—once it was deemed a money-losing proposition—it was shut down.

In the year 2010, dog cloning, a technology developed primarily for the benefit of the very rich, was being offered by only one company—in the city of Seoul, on the planet of Earth.

There, men and women in white lab coats and hairnets sit at the helm of micromanipulators, using joysticks to transfer cells into eggs, zappers to apply the electric jolt needed to make those cells merge, and surgery to implant the fused future embryos into the uteri of surrogate mother dogs. Thousands of years after they made the transition from wild to domesticated, less than a century after making the transition from laborer to best friend, just a few short decades after they became, in effect, our roommates, dogs became "cloneable"—a technology that quickly made its own transition: from the laboratory to the marketplace.

A decade into the twenty-first century, dog cloning appeared to be an industry with at least something of a future.

By the beginning of 2010, nearly twenty canine clones—between those produced through BioArts and RNL Bio—had been delivered to paying customers who wanted a copy of their pets. While BioArts made its exit from the dog-cloning business in 2009, the Seoul-based RNL Bio was expanding. RNL has announced plans to build a $5 million facility, capable of pumping out a thousand dog clones a year. Company officials predicted RNL would be cloning a hundred dogs in 2010, and up to five hundred by 2012. The cloning facility was to be built outside Seoul, in Yongin—on the same mountainside, in fact, as Woo Suk Hwang's Sooam Institute.

Lou Hawthorne's withdrawal from the cloning market didn't leave Hwang idle. Although the government had rejected his request in 2008 to return to experimenting with human embryos—"the person in charge of this research has some ethical flaws," a panel noted—Hwang was teaming up with local governments. In August 2009 he finalized a deal to collaborate with the province of Gyeonggi to clone transgenic animals for the purpose of growing organs for transplantation to humans. Six more provincial governments had reportedly contacted him to collaborate on biotech projects. He had submitted several research papers related to his cloning process that were under review at *Nature* magazine. And he was again joined in his lab by Taeyoung Shin, once his student, later a key scientist in the Missyplicity Project and vice president of BioArts.

Both BioArts and RNL by then had branched into offering more services for humans. BioArts, though shutting down its dog-cloning arm, Encore Pet Science, was going through a period of reorganization, marketing DNA parentage tests to Chinese citizens and selling the GoldiLox, a container for shipping tissue samples developed by one of its subsidiaries, replacing what Hawthorne said had previously been the industry standard—an ice chest.

RNL had moved on to banking the stem cells of human customers, each of whom paid $3,500 for the service in hopes that the cells might someday cure them of diseases, including diabetes, heart problems, cancer, or even spinal-cord injury. By fall of 2009, twenty people, mostly Korean-Americans, had gone to Korea to have their stem cells gathered prior to having them reinjected in countries in which that is not outlawed. RNL established clinics in Germany and Japan for that purpose, and was looking into opening one in Tijuana, Mexico. RNL had also introduced a line of stem-cell cosmetics through an office established in Beverly Hills.

Only in Korea were dogs being cloned—mostly for American customers and mostly behind closed doors, with no official oversight and without all the special animal-welfare precautions Hawthorne had prescribed during the Missyplicity Project. Dogs used for cloning in Korea weren't going on to adoptive homes—just, if lucky, back to the farms or suppliers they came from.

Despite conviction on criminal charges, Hwang at Sooam and Byeong Chun Lee at SNU continued their research mostly unfettered by rules and regulations. Internationally, with some new doors opening, cloning research continued its quiet march forward.

In March 2009, President Obama lifted the President Bush–imposed restrictions on federally funded embryonic stem-cell research in the United States. In May, the Louisiana Senate approved legislation that would make it illegal to combine human and animal cells to create "human-animal hybrids." Conviction would result in ten years in prison for any researcher who attempted it. The law didn't stipulate what would be done with any resulting "manimals." In June, scientists at India's National Dairy Research Institute produced the world's first surviving buffalo clone. At the end of the month, Ian Wilmut, the "creator" of Dolly the sheep, was knighted by the queen for his career achievements. In July, the European Union decided meat and milk from clones should be allowed on store shelves, and representatives of Clonaid, the company formed by religious leader Raël, which claims to have cloned dozens of humans, issued a coyly worded press release saying that while they had met with Michael Jackson before his death, privacy concerns prevented them from saying whether they were cloning him.

Simon Brodie, the former operator of ForeverPet, who later established a nationwide dog-leasing company, saw FlexPetz go on hiatus after the Boston City Council passed a law banning pet rentals. By 2009, he had moved to Big Sky, Montana, where, under the name Simon Carradan, he was selling what have been called the world's most expensive skis.

Meanwhile, down in Texas, Ralph Fisher was getting a mount made of Second Chance for display at his ranch in LaGrange, where the pelt of the original Chance remains stashed in a closet. CC, the first cat clone, was continuing to enjoy her family and condo, not far from Texas A&M, which still held the record for having cloned the most species. And in Austin, Jessica Harrison, who tended the dogs during the Missyplicity Project, had decided against cloning her Australian shepherd, Hunter.

"Obviously they're going to look the same, but if they don't act the same, then what's the point?" she said. "I will tell you that after going to meet the clone [of Missy], I decided I don't want to do that. . . . It would take away from the uniqueness of him, and I'd have too high expectations for the puppy, who would have these super-huge shoes to fill."

In Florida, Edgar and Nina Otto—among the winning bidders for a BioArts cloning—are enjoying the antics of Lancelot Encore, the clone of their dead yellow Lab Lancelot, at their twelve-acre estate in Boca Raton.

Edgar Otto, Jr., the son of a cofounder of NASCAR, who made a fortune in thermal food trays for institutional use, spent $155,000 on the cloning. They'd banked Lancelot's cells with GS&C five years earlier, after he came down with nose cancer.

Otto, who still drives race cars in his eighties, owns a nationwide chain of wound treatment centers and is the inventor of URAssist—named You're-In Control when it was still in the prototype stage. It's a urine collection and containment system that attaches to bed or wheelchair for men and women with bladder control or mobility problems.

The Ottos, generous donors to their local humane society, say Lancelot Encore, the first clone to be delivered by BioArts, was, for them, a worthy investment. "We don't own an airplane. We don't own a boat. We don't own lots of things that cost a lot more than a dog that we loved and missed," Edgar Otto said. "I would say that anybody that's lost a dog can understand our motivation. Anybody who says 'shame on you' has never really had, or earned, the love of a dog."

While he isn't an exact duplicate of the original, behavior-wise the clone is exhibiting one of Lancelot's endearing traits. Invariably, when he would lie down, Lancelot would cross one paw over the other—not an entirely unique behavior among dogs but a telling one to Otto. "That's one of the things he does. Lancey right now is crossing his paws 60 percent of the time."

The Ottos have also noticed that Lancelot Encore—now just called "Lancey"—likes to spend time next to a bush they planted in the spot where the original Lancelot died in 2008. Lancelot is not buried there; it's just where he drew his final breath. But his clone, the Ottos say, seems drawn to the spot, unlike any of their nine other dogs.

BioArts delivered all four other clones to winning bidders in its online cloning auction before it shut down its pet-cloning division in the fall of 2009.

By then, RNL Bio was working on at least eight more commercial clonings—including that of a shih tzu for a Korean-American client, a German shepherd for a Canadian living in Thailand, and another for a prominent Chinese politician the company declined to identify. They had delivered the clone of a Pekingese named Snow to a man they described as a "global traveler of Middle Eastern descent." Apparently, the customer was pleased; after just a short time with his clone, he put in an order for five more copies. RNL was also working on cloning a dog under a contract with PerPETuate, the American company that continues to bank pet genes.

Yet another RNL cloning customer, Peter Austin Onruang, a former street-gang member who now sells spy gear from a shop in Hollywood, was waiting for the clones of his two deceased Yorkshire terrier–Schnauzer mixes, Wolfie and Bubble.

Onruang, who transformed a bedroom business selling car alarms over the Internet into Wolfcom Enterprises, named after Wolfie, was ready to sell both his Hollywood shops—a second one sells paintball equipment—to pay for the clonings after Wolfie's death in 2009. "I'd be poor again just to be with them," he said. "I was special because of Wolfie and Bubble. Once they're gone, I'm not special. I'm just a normal guy."

Wolfie accompanied him to work each day, and he credits the dog with turning his life around. After the dogs came into his life, when he was twenty-four, he developed a sense of responsibility. "Every time I thought about doing something illegal, I thought if I was in jail I wouldn't be with them," he said. Both dogs shared his bed, Bubble perched on the pillow above his head, Wolfie sleeping alongside him. Often, he would wake up and see Wolfie staring at him. Twice, he said, Wolfie stopped him from committing suicide. "A dog instills in us something we don't have, or something we do have but don't use," he said. "It's love. It's compassion. It's forgiveness. Dogs are quick to forgive. They don't lie. They don't cheat."

By the time of Wolfie's final birthday celebration, in 2009, the dog was in sad shape. Onruang closed up shop so his staff, all of whom donned party hats, could attend. Wolfie, blind in one eye, getting periodic doses of oxygen and subsisting on food that Onruang would run through a blender and push into her mouth via syringe, watched it all motionlessly, wrapped in a turquoise blanket. She was totally paralyzed by then.

She died the next day. "I hadn't been alone for fifteen years. It was very hard," Onruang said. "I went to the animal shelter and looked at a couple of dogs, but I just couldn't do it. They were cute and everything, but it would be wrong." Instead, he pursued cloning.

Onruang admitted to having some mixed feelings about cloning after he returned from a visit to RNL and Seoul National University in May. "I met the dog that glows in the dark. I met Snuppy. I met the main doctor. They were all very nice." But learning about Snuppy's living conditions and hearing the whining moans the world's first dog clone is prone to emitting left Onruang with a sense of discomfort. "It was almost like he was trying to tell me something. It's almost like he's crying. He's caged, he has no life. That's the part of cloning that bothers me—that and the possibility of deaths and deformities and dogs being kept for these experiments."

The possibility of getting his dogs back—he thinks it is possible that the souls of his old dogs will enter the bodies of his new ones—outweighs those concerns.

"People say they will just be a copy and not have the same soul," he said. "I accept that, but I also think if I raise them the same way, then the clones will grow up to be similar in personality, and the spiritual part of me hopes that maybe the souls of Wolfie and Bubble will go right back in. Spiritually, we don't know anything about our universe. How do we know that when these dogs are born, the souls of Wolfie and Bubble won't go back in there? They probably could. They probably would. They probably will."

Not far away, at the home of James Symington, the ex–police officer was training all five clones of his German shepherd, Trakr, to be search-and-rescue dogs, trying to sell his dog's story as a book or movie, and had established the Team Trakr Foundation, described as "an international humanitarian organization dedicated to providing and advancing the availability of elite, self-contained K9 search-and-rescue teams in the United States and around the world." Its mission is to support, train, and deploy the clones of Trakr when humans are lost or trapped in disasters.

While the pieces seemed to be falling into place for Symington, Joyce Bernann McKinney and her five clones continued to bounce in and out of turmoil.

Six months after she managed to get them home from Korea, all five Booger clones had experienced stomach viruses, and a few have had more serious problems. Ra suffered an anal prolapse not long after getting home and required surgery. He has also shown signs of epilepsy, having mild seizures and balance problems. He bites at flies that aren't there. Mack, named in honor of her family—and clearly the alpha dog of the group—has what McKinney described as an eating disorder, often refusing to take food, and he has behaved aggressively with some of the other clones.

Many of the dogs McKinney already had—five of them—hadn't taken well to the new family members. Three of her rescued pit bulls have shown aggression toward the clones, and one, a rescued pit bull named Baby, bit her housemate Brown so severely that, a year later, he was still undergoing physical therapy to regain full use of his right arm.

Between her repeated trips to ferry the dogs home, hotel stays, taxi fares, and paying for hotel-room damage, she had ended up spending $22,000 of her father's money for her trip to Korea. Once back home in Riverside, she found that Brown had not built the fence—chain-link, not white-picket—he had promised to put up during her absence. Because of her fears that her existing dogs would hurt the clones, and the clones' tendency to fight one another, they'd remained mostly crated, indoors, since her return.

"I promised them in Korea that I would free them and that they'd never see a cage again. But I had to put them in a cage again, just to keep the other dogs from attacking them or killing them or something. They've been in cages now since November and I think they've gone cage-crazy."

In July, RNL's Jin Han Hong attempted to visit McKinney and the dogs, but was stood up. The next day, McKinney, frustrated with what she saw as Brown's inability to get things accomplished around the house, called the county mental-health office, saying he needed help dealing with what she called his Alzheimer's. The mental-health office asked the sheriff's department to make a visit. Upon seeing the condition of the home—urine-soaked carpets, feces, hoarded items, and general disarray—deputies declared the house a health hazard. Witnessing the couples' arguments and hearing Brown claim that McKinney had used his credit card without his permission, they arrested her and jailed her on suspicion of elder abuse. The charge was later dismissed for lack of evidence, but as a result of her jailing and the home's condition, all ten of her dogs—the five clones included—were swept up by Animal Control of Riverside County.

Released from jail the next day, McKinney went to the pound and rescued the world's first commercially cloned dogs, and her other dogs as well. She put her non-cloned dogs in kennels and began searching for a place to stay. By July 2009, McKinney was once again bouncing among motels with her crated clones. They didn't get along with one another; and if she let more than one out of his crate at a time, a fight would often ensue.

Brown invited her to return to the house, on the condition the dogs stay outside, and McKinney—after several weeks—agreed. She moved back in, and the fence got built. Life in Riverside became somewhat calmer, if not the idyllic ending RNL painted.

"For a great deal of time she drove all of us crazy," admitted Jeong Chan Ra, CEO of RNL. "But she has found peace. She was so happy just to have her dogs and have that reconnection. In the aftermath, looking at all we went through, and finding about her past and having to deal with it, media-wise, it was still a worthwhile experience for us. To give somebody that kind of happiness is priceless."

In truth, McKinney was up and down—reveling in her clones one day, questioning her decision to clone Booger the next.

"I have to say that cloning ruined my life, you know, it ruined my life," McKinney said at one point. "I'm in fear constantly that I'll lose one of them, like I did my precious Booger, and it would be like him dying all over again. Sometimes when Ra is having a seizure and I'm holding him in my arms and his eyes are rolling back in his head, I ask myself: did I do the right thing by cloning? Am I a bad person? Was I selfish, like all those chat-room people said I was?

"All I was trying to do was have my Booger back," she said in one phone call, between sobs. "I was just trying to clone my friend."

It was far from the happily-ever-after ending she envisioned. In fairy tales, life's limited duration can be ignored. Dragons can be slain. Mice can become horses. A beauty can fall in love with a beast and, without surgery, micromanipulation, electric shock, or even handcuffs—with only a postmortem smooch—watch him transform into her handsome prince. McKinney found that, in a world where the "ever after" part has become, in a way, achievable, there's still no guarantee on "happily"—not even with your beloved dog returned fivefold.

In Northern California, meanwhile, Mira, the first of the Missy clones, continued to thrive at the Marin County home of Lou Hawthorne and his son, Skye. While Skye's grandmother didn't want her clone—MissyToo ended up at Sperling's Arizona estate—Mira was living a happy dog's life, making daily visits to the park down the street.

There, as the sun dipped behind the hills, father and son were walking down a winding path on a fall evening, their dog—with her one ear floppy, her other ear straight—following a few steps behind, when Skye decided the time was right for a "slightly scary" story.

"It was a monster, right?" his father says.

"Not a real big creepy monster, but sort of a bit out of the ordinary. I was playing up here with the dogs, and I went down there to go to the bathroom in the woods," Skye says, pointing to a ravine to the right. "I heard a rustling in the trees and I didn't know what it was. Then I heard it again. I just caught a glimpse from the corner of my eye of a huuuuuuge doglike thing. . . . But it was probably not a dog. . . . Dogs don't run that fast."

Just then, his own dog, Mira, all but proves him wrong. She spots a deer on the ridge above the road. For a few frozen seconds, they stare at each other. Then Mira takes off—every bit as speedy and fearless as the dog she was cloned from. She bounds up a steep incline, kicking up turf and gravel and disappearing into the dusk. Father and son wait patiently, assuming a similar arms-crossed stance. Clearly, this has happened before.

Skye uses the opportunity to finish the story of his beastly encounter—one that ends leaving the listener hanging. Is it fact or fiction? And just what was the creature? Skye seems to enjoy the mystery and, maybe even more, the control he wields in weaving it. "What was it?" he says, repeating the question and allowing for a dramatic pause. "I guess we'll never know."

Minutes later, a panting Mira—having been either outwitted or outrun—reappears. She slides back down the incline, brushes past the pant leg of Lou Hawthorne, then trots up to Skye. The boy ruffles her neck fur, holds her head in his hands, and looks into her eyes.

"You're a clone, Mira," he says.

"You're a clone. You're a clone."

Acknowledgments

The author wishes to thank his editor, Rachel Holtzman, at Avery, for her patience and guidance and for making this book much better; his agent, Neeti Madan, of Sterling Lord Literistic, Inc., for finding the book a home; and the president of Sterling Lord Literistic, Philippa Brophy, without whose faith and support the book wouldn't exist. Special thanks as well to Rona Kim, whose assistance as an interpreter and guide in South Korea was invaluable.

Index

Abandoned dogs, 14
Abbey, Edward, 44
Abnormalities, in cloned dogs, 280–81
Ace (author's dog), 7
Actor, Symington as, 42
Adaptations of dogs, 14–15
Afghan hounds, 181
After Dolly: The Promise and Perils of Cloning (Wilmot), 126, 139
Against All Odds (John Sperling), 51–52
Air Liquid Group, 75–76
Aliens Adored (Palmer), 74
Allerca, 157
"Allergy-free" cats, 150
Alta Genetics, 135
American Anti-Vivisection Society, 154
  and cloning of animals, 159–60, 216
  and "Golden Clone Giveaway," 265
American Breeders Service, 135
American Kennel Club, 15
American Society for the Prevention of Cruelty to Animals (ASPCA), 216
Americans, and dogs, 22–24
  hero dogs, 38
Anderson, Kirk, 241–45, 253
Animal cloning, 6, 43, 58, 104–13, 127–39, 217–19
  first human-assisted, 128
  museum exhibit, 147
  John Sperling and, 281–82
  See also Dogs, cloning of; Pets, cloning of
Animal control agencies, 14
Animal Farm (Orwell), 36
Animal welfare groups
  and cloning, 154, 216, 218–19, 232
    commercial, 215
  Korea, 219–20
  and online cloning auction, 265
  and pet rental, 159
Animalhelp.com, 103
Apocalypse (Raëlian newsletter), 73
Apollo Group, 53
Aristotle, 130
Ashes, of cremated pets, 118–19

Index

ASPCA (American Society for the Prevention of Cruelty to Animals), 216
Auction, online, of cloning, 13, 264, 272
Audubon Center for Research of Endangered Species, 146–47
Auto Pop magazine, 72
Auto racing, Raël and, 155
Automobile, buried on Earth Day, 50–51
Avner, David, 157–58

Baba Ganoush (cat clone), 152–53
Baby (pit bull), 294
Baby Louise. *See* Brown, Louise
Baegun (Korean monk), 143
Balto (Alaskan hero dog), 40–41
Bass, Edward P., 90
Beagles, cloned, 195
  transgenic clones, 225, 228
Bear (search-and-rescue dog), 101, 269–70
Bear: Heart of a Hero (Shields), 270
Bear Search and Rescue Foundation, 269
Beauty queen, and Mormon missionary, 241–47. *See also* McKinney, Joyce Bernann
Bee Gees, 237, 241, 246
Benji (poodle), 146
  clone of, 195
Bergh, Henry, 216–17
Bessoff, Heather J., 82
"Best Friends Again" program, 207
"Best Friends Again" website, 211
BioArts International, 13, 191, 207–9, 214, 282, 286, 287, 288
  cloning auction, 290, 291
  cloning of Trakr, 215
  commercial cloning, 264–72
  and Woo Suk Hwang, 276
  and media, 270
  patents held by, 274
  withdrawal from dog cloning, 278–80
Bioserv, 77
Biosphere 2, 90–91
Biotech research, dogs used in, 221
Blue Dragon (Sapsaree), 224–25
Boer goat, cloning of, 105
Boisselier, Brigitte, 75–76, 77–79, 155
Bona (first female cloned dog), 195
Bonnardot, Alfred, 122
Bono, Sonny, 60
Booger (pit bull), 16, 29–35, 114, 215, 249
  clones of, 230–35, 248–50, 256–62
    health problems, 294
  death of, 184–85
Boredom, John Sperling and, 47–48
Bowen, Ray, 47
Boys from Brazil, The (movie), 239, 240
Brahman bulls, 59–60
Brain Korea 21, 150
Brave New World (Huxley), 131, 133, 263
Break Through Prayer: The Secret of Receiving What You Need from God (Cymbala), 97
Breeding of dogs, 15–16
BRIC (Biological Research Information Center) website, 199
Briggs, Robert, 132
Brodie, Simon, 156–59, 289
Bromhall, Derek, 240
Brookside Zoo (Cleveland), 40
Brown, Elliott, 33, 114, 256, 259, 260, 294–96
Brown, Louise (test-tube baby), 128, 238, 239, 247
BTX Electro Cell Manipulator, 65, 81
Bubble (Onruang's dog), 292–93
Buchanan, Brian, 98–99
Buddha, 140, 141
Buffalo, clone of, 289
Bull 86 (Angus bull), 105
Bullet (Roy Rogers's dog), 122
Bulls, tame, 58
Bush, George W., 180
Butterflies, release of, 249–50, 258
Buttermilk (Dale Evans's horse), 122

Calagan, Chris, 115, 121
Calagan, Sandy, 115
Calico cats, 108
  Naomi (Calagans' cat), 115–17
  Rainbow (donor cat), 108–9, 111
Call of the Wild, The (London), 93
Camel, cloning of, 277
Campbell, Keith, 137
Cantrell, Bertie, 255
Capshaw, Kate, 60
Carbon Copy (CC, cat clone), 106, 109–12, 153, 290

Index

CARE (Coexistence of Animal Rights on Earth), 219
Carlson, Ben, 178
Carradan, Simon (Brodie), 289. *See also* Brodie, Simon
Carter, Lynda, 242
Cat in the Hat Comes Back, The (Dr. Seuss), 104
Catholic Church, and cloning research, 132
Cats, cloned, 106–12, 152–53, 162–63, 218, 290
  hypoallergenic, 157–58
Cattle, cloning of, 135, 146
CC (Carbon Copy, cat clone), 106, 109–12, 153, 290
Cell (biology journal), 133
Celler. *See* Vorilhon, Claude
Center for Genetics and Society, 266
Chance (Brahman bull), 57–65, 171, 173
  clone of, 65–69, 105
  pelt of, 290
Chaser (drug-detecting dog), 225
Chicago Museum of Science and Industry, 147
Children, lessons from dogs, 2
"Chimeras," in taxidermy, 122
Chinese Academy of Sciences, 227
Cho, Sean, 258, 259–60
Choe, Yun Ui, 140–41
Chromosomes, gender-related, 108
Cities, dogs in, 23–24
Clients of dog-cloning companies, 275
Clinton, Bill, 138
Clonaid, 70–71, 75–76, 155, 289
Clonapet, 70–71, 76, 154–55
"Clone," first use of word, 133
Cloned animals, 66
  dogs, 287
  See also Pets; *specific animals and animal names*
Cloning of animals. *See* Animal cloning
Cloning and Stem Cells, 277
Cloudhoppers, 156–57
Coexistence of Animal Rights on Earth (CARE), 219
Commercial cloning
  of dogs, 17–18, 43, 189, 207–10, 214–16, 291–92
    competition in, 273–84
    RNL Bio and, 189, 226–27, 232–36
  of pets, 150–51, 177–78, 286–88
    Koreans and, 206–10
Congressional hearing, Raël and, 155–56
Corporate interests, public universities and, 107–8
Cosmetics, stem-cell, 288
Cow, first cloned, 135
Cremation of pets, 118
Crick, Francis, 155
Criminal charges
  against Woo Suk Hwang, 275–76, 277, 283–84
  against Lee, 275–76, 283–84
  against McKinney, 252–54
Culture, Korean, and dog cloning, 142–43
Cushman, Rick, 98–99
Customers of dog-cloning services, 275
Cyagra, 82
Cymbala, Jim, 97

"Daedalus, or Science and the Future" (Haldane), 133
Daily Mirror (UK), 247
Dallas Morning News, The, and John Sperling, 54
Darrow, Clarence, 9
Death, children's experience with, 2–4
Deer, cloned, 127, 187–88
  fluorescent, 157
Deformities, in cloned dogs, 280–81
Denniston, Richard, 82
Dewey (deer clone), 188
DNA banking, 16, 83, 154, 187–88
  of Booger, 185–86
DNA parentage tests, 288
Dog market, Korea, 142–43
Dogs, 2, 4–7, 13–16, 22–24, 217
  availability of, in Korea, 280
  cloning of, 16–19, 20–22, 25–26, 105–6, 111–12, 217–19
    commercial, 17–18, 43, 189, 207–10, 214–16, 226–27, 232–36, 291–92
    competition in, 273–84
    Genetic Savings & Clone and, 153
    Lou Hawthorne's views on, 279–81
    health problems, 294

Dogs *(cont.)*
in Korea, 141–51, 174–83, 193–95, 206–7, 223–24, 271–72, 286–88, 291–92
online auction of, 13, 264, 272
process of, 64–65, 148–50, 175–76, 270–71
John Sperling and, 54
at Texas A&M, 80–81
as food in Korea, 219–21
as heroes, 38–41
hypoallergenic, 158
memorializing of, 115
number, in U.S., 24
DogsintheNews.com, 103
Dolly (sheep clone), 22, 125–27, 128, 137–39
Dolly patents, 274
Donadio, Dan, 96
Driesch, Hans, 128, 129
Drugs, war on, John Sperling's views on, 53

Earth Day, 50
Ectogenesis, 133
Education, online, 52–53
Eggs, human, sources of, 197–98, 200, 201
86 Squared (bull clone), 105
Elohim (planet), Vorilhon and, 72–75
Embryonic Development and Induction (Spemann), 132
Embryonic stem-cell research, 289
Encore Pet Science, 272, 276, 278, 286
closing of, 281
Endangered species, preservation of, 227
Esposito, Billy, 100
European Union, and clones, 289
Evening Argus (UK), 156
Exeter Life Sciences, 208
"Experiments on Plant Hybrids" (Mendel), 130
Extraordinary Service to Humanity Award, 103, 269
Extraterrestrial experience, Vorilhon and, 72–74

Family members, dogs as, 23–24
Fashion accessories, dogs as, 14
Fear Itself (Joan Hawthorne), 212
Females, X chromosomes, 108
Fields, W. C., 174
Fighting dogs, 14, 28
Films, Lou Hawthorne and, 90–91
Finnegan Forcefield (cat clone), 163
First, Neal, 135
"First Nine Lives Extravaganza," 152
Fisher, Ralph, 56–69, 166–73, 289–90
Fisher, Sandra, 60, 61, 64, 66, 67, 69, 168–69, 171, 172
FlexPetz, 158–59, 289
Fluorescent transgenic dogs, 225, 228, 272
Food, dogs as, 142–43, 219–21, 280
Food and Drug Administration, 78–79, 217
ForeverPet, 116, 156, 157, 289
Fox Chase Cancer Center, 132
Frankenstein (Shelley), 184
Frankie Forcefield (Lynch's cat), 162–63
Fraud, 154–58
Woo Suk Hwang and, 193–94, 199, 201–2, 276, 283
Freeze-drying, 116–18, 120
Frogs, cloning experiments on, 132–33
Funding of cloning research, 81–82, 107, 134–35, 143, 150
Funerals, for pets, 118

Galione, Bob, 99–101
Gene banking, 16, 83–84, 272
Genesis (Bible), 125, 127–28
Genetic engineering, Wilmut and, 136
Genetic Savings & Clone (GS&C), 82–84, 127, 141, 142–43, 153, 154, 163–65, 178, 186–87, 206–7, 286
cat clone, 162–63
criticisms of, 160–61
"First Nine Lives Extravaganza," 152–54
Shin and, 148
Texas A&M and, 106–10, 111–13
Geneticas Life Sciences, 157
Gere, Richard, 40
Germ-Plasm, The: A Theory of Heredity (Weismann), 129
Geron Corporation, 208
Gibb, Maurice, 246
Gillespie, Ron D., 82
Glass, Ira, 167–68
"Golden Clone Giveaway," 13, 215, 264–67
GoldiLox, 288
Goodall, Jane, 103

Index

Grace & Co., 135
Gracie (Oprah Winfrey's dog), 119
Graeber, Charles, 81
Granada Corporation, 135
Greene, Joanne, 266
Ground Zero, work at, 95–96
GS&C. *See* Genetic Savings & Clone
Gurdon, John, 132–33
Gutenberg, Johannes, 141
Guzman, Genelle, 95, 96–99, 269

Ha, Ji Hong, 223
Ha, Sung Jin, 222–23
Hachiko (Japanese hero dog), 39–40
Haetae (Korean sculpture), 39
Haldane, J. B. S., 133
Halifax, Nova Scotia, K-9 unit, 37
Halifax *Chronicle Herald*, 102
Hall, Joe, 94, 102
Han, Kook Il, 223–25
Hanwha L&C Corp., 276
Harrison, Jessica, 85, 266, 290
Hawthorne, Joan, 21, 22, 49–50, 54, 92
  and children, 85–89, 91
  and Missy, 24–26
  and Missy's clone, 212–13
Hawthorne, Lou, 12–13, 21, 22, 25–26, 44–46, 54, 85–92, 186–87, 297
  and cloned cat, 109–10, 153
  and cloning of dog, 105
  and criticisms, 160–61
  and Encore Pet Science, 272
  and "Golden Clone Giveaway," 264–67
  and GS&C, 164
  and Woo Suk Hwang, 206, 207–9, 268–69
  and Korean commercial cloning, 208–10
  McKinney and, 260–61, 191
  and media, 270
  and Missyplicity project, 81–82
  and Missy's clone, 206, 210–11
  and pet cloning, 82–84, 267–68
  and RNL Bio, 274–75
  and Second Chance, 67
  and Trakr clones, 268
  Westhusin and, 113
  withdrawal from dog cloning, 278–79, 282
Hawthorne, Skye, 211, 297–98
Hell's Buddhas (film project), 91
Helmsley, Leona, 24
Heroes, 36
  cloning of, 43
  dogs as, 38–41
    Trakr as, 38
Hong, Jin Han, 189, 190, 251, 257, 295
Hope (dog clone), 195
Hoppe, Peter, 133
Horse, first cloning of, 127
Huggable Urns, 118–19
Human-animal hybrids, 289
Human organs, from transgenic animals, 287–88
Human stem cells, banking of, 288
Humane Society of the United States, and pet cloning, 215–16
Humans
  cloning of, 17, 19, 43, 71, 74–79, 195–96
    Korean project, 180
    Raël and, 155
    stem-cell lines, 176
    Wilmut and, 139
  and dogs, 15–16, 22–23
Hunt, Mark, 76–79
Huxley, Aldous, 20
  Brave New World, 131, 133, 263
Hwang, Cheol Yong, 180–82, 183
Hwang, Woo Suk, 144–48, 175, 176–77, 180, 193–204, 214, 266, 270–71, 278, 287–88, 289
  criminal charges against, 275–76, 277, 283–84
  and dog cloning, 207, 210, 227–28
  Lou Hawthorne and, 206–9, 268–69
  lawsuit against, 282–83
  RNL Bio and, 275

Iditarod sled-dog race, 41
Illmensee, Karl, 133–34
In His Image: The Cloning of a Man (Rorvik), 239–40
Inbreeding of dogs, 15–16
India, buffalo cloned in, 289
Industry, dog cloning as, 287–88
Infigen, 164
Insuraclone, 76

Index

Internet
and hero dogs, 41
and Woo Suk Hwang, 198–99
and memorialization of pets, 119–20
and pet cloning, 233
and Trakr, 103
University of Phoenix and, 52–53
Ispig (transgenic pig clone), 276

Jabari GD (hypoallergenic dog), 158
Jackalopes, 122
Jackson, Michael, Clonaid and, 289
Jang, Goo, 179
Japan
hero dog, 39–40
and Korean dogs, 222
Jazz (test-tube wildcat), 146
Jeong, Hae Jun, 203
Jikji (Buddhist text), 143
Judd, Ashley, 60

Kahless (clone of Missy), 211
Kang, Won Rae, 199
Kim, Min Kyu, 142–43, 148–51, 177, 194
Kim, Sun Jong, 200
Kimble, George, 40
King, Stephen, 205
King, Thomas, 132
Korea. *See* South Korea
Korea Herald, The, 284
Kraemer, Duane, 110–11
Kronos (antiaging clinic), 53

Laboratory dogs, 84–85
Lachini, Alexandra, 118–19
Lancelot Encore (dog clone), 290–91
Lassie Come Home (movie), 1
Late Night with David Letterman, 60
Lauria, Joe, 78
Lawrence, Candida (Joan Hawthorne), 49, 86. *See also* Hawthorne, Joan
Lawsuits, 254–55, 275–76
Lazaron Biotechnologies, 82
Leary, Sue, 154
Lee, Byeong Chun, 144, 145–48, 151, 177, 179, 181, 183, 193–94, 202, 210, 214, 215, 224–27, 289,
and Booger clones, 232
criminal charges against, 275–76, 284
McKinney and, 189
Let's Welcome Our Fathers from Space (Raël), 155
Levin, Ira, 239
Lewis, Peter, 53
Liebe (Joan Hawthorne's dog), 25
Life, of dog, prolonging of, 5–6
Lifestyle Pets, 158
Lightning-strike.com, 120
Little Gizmo (cat clone), 153–54, 159
owner of, 160–61
Little Prince, The (Saint-Exupéry), 285
Livestock, cloning of, 134–35
Locusts, The (movie), 60
London, Jack, 93
Los Angeles Times, 177
Lynch, Liam, 161–63, 266

Mack (clone of Booger), 294
Maggie (mummified poodle), 123–24
Males, X chromosomes, 108
Mammoths, cloning attempts, 127, 277
Mangold, Hilde, 131
Marine (cancer-detecting dog), 225–26
clones of, 235
Marketing of dog cloning, 16–17. *See also* Commercial cloning
Maverick (car), buried on Earth Day, 50–51
May, Keith, 242–43, 245, 246–47
McGrath, James, 134
McKinney, Joyce Bernann, 9–11, 16, 27–35, 114, 207, 215, 229–36, 237–47, 250–56, 269, 294–97
and Booger's death, 184–88
and clones of Booger, 188–91, 248–50, 256–62
Media
and cloned bull, 64–65, 167–70, 173
and cloned sheep, 138–39
and dog clone, 175, 176–77
of Trakr, 264–65, 269
and dog rental, 159
and hero dogs, 41
and Woo Suk Hwang, 196–98, 200–201
and McKinney, 230, 233–36, 252, 255
Mormon and beauty-queen story, 244–47
and Missyplicity Project, 81–82

Index

and 9/11, 101
and pit bulls, 28
Medical advances from cloning, 139
from cloned stem-cell lines, 196
dog cloning and, 150, 177, 226
Megan (Boer goat clone), 105
Memorialization of pets, 115, 119–20
Mendel, Gregor, 129–30
Menu, Su, 123–24
Mice, cloning of, 133–34
Miller-Spiegel, Crystal, 218–19, 265
Mira (clone of Missy), 210–11, 212, 297–98
Miss World contests, 242
Missy (Joan Hawthorne's dog), 21, 24–26, 44, 46, 54, 205, 212–13
clones of, 206, 210–12, 272
cloning of, 55, 268–69
Lou Hawthorne and, 92
MissyToo (clone of Missy), 212–13, 297
"Missyplicity Code of Bioethics," 84–85
Missyplicity Project, 46–47, 62, 83, 106
funding for, 81–82
Mitochondrial DNA, 212
Monkey, cloning attempt, 127
Moon, Keith, 246
Moran Market, Korea, 143, 220–21
Mormon Church, McKinney and, 242, 243
Mormon missionary, beauty queen and, 241–47. *See also* McKinney, Joyce Bernann
Movable type, invention of, 140–41
Mule, first cloning of, 127
Mummification, 123–24

Names of dogs, 23
Naomi (calico cat), 115–17
National Cancer Institute, 132
National Dairy Research Institute (India), 289
National Fallen Firefighters Memorial, 100
Nature, 134–35, 288
and dog cloning, 177–78
Nature Biotechnology, and cloned cat, 111
New Scientist, 160
New York Times, The
on Biosphere 2, 90
on cloned animals, 21–22, 138–39, 177
9/11, 41, 94–103
1978, 238–41

Newman, Paul, 155
Nitro, West Virginia, 77
Noah's Ark, 73, 127–28
Nome, Alaska, diphtheria epidemic, 40

Obama, Barack, 289
Online dog-cloning auction, 13, 264, 272
Onruang, Peter Austin, 292–94
Organ transplants, Koreans and, 180
Organs, human, cloning of, 287–88
Orwell, George, 36
Oscar (buzzard), 60–61
Otto, Edgar and Nina, 290–91
Ovadonation.or.kr, 198
Overpopulation of pets, cloning and, 282
Ovulaid, 76
Ownership of pets, 4–5, 121

Pacelle, Wayne, 215–16
Palmer, Susan J., 75
Pampering of dogs, 13–14, 23–24
Park, Soyoun, 219–20, 221, 227
PD Su-cheop (Producer's Notebook; Korean TV show), 197–98, 200–201
Peace (dog clone), 195
Peaches (cat clone), 153
Permanent Body Preservation System, 124
Perpetual Pet, 116–17, 120
PerPETuate, 82, 116, 292
Pet owners, 43, 116
Pet rocks, 240
Petloss.com, 120
Pets
cloning of, 58, 82, 127, 150–51
Clonaid and, 1, 76
commercial, 226
Lou Hawthorne and, 267–68
opposition to, 154, 233
Seoul National University and, 177–78
Wilmut's views on, 139
freeze-dried, 117–18, 120
Hawthornes and, 89
mummification of, 123–24
rental of, 289
Pet Sematary (King), 205
"Pharming," 136
Pigs, cloned, 105, 276
Pit bulls, 28–29
Police dog, Trakr as, 37–38, 41

Index

Population, of dogs, in U.S., 24
Posthumous portraiture of pets, 121
Prather, Randall, 135
Professors and academics, John Sperling and, 49
Proverbs (Bible), 166
Public universities, and corporate interests, 107–8
Purebred dogs, 15–16
  Koreans and, 219
Puy de Lassolas (France), 71–72

Ra (clone of Booger), 294
Ra, Jeong Chan, 189, 226–27, 235, 274
  and McKinney, 296
Raël. *See* Vorilhon, Claude
Raëlian Movement, 71, 73–75
Rainbow (calico cat), 108–9, 111
Ralph Fisher's Photo Animals, 57, 61
Rat, first cloning of, 127
Rather, Dan, 60
Rebel with a Cause (John Sperling), 48, 281–82
Reeve, Christopher, 196
Reggio, Brett, 82
Reincarnation, cloning and, 127
Relationships, human
  with dogs, 4–6, 22–24
    in Korea, 221–22
    McKinney and, 28
    Symington, with Trakr, 37–38
  John Sperling and, 54, 92
Rental of pets, 289
RePet, 285–86
Research
  cloning, funding of, 20–21, 53–55, 81–82, 107, 134–35, 143, 150
  embryonic stem-cell, in U.S., 289
Reveille (Texas A&M mascot), 45, 55
"Rites of Transference," 124
RNL Bio, 189–90, 207–10, 214–15, 287, 288
  and commercial pet cloning, 206, 226–27, 274–75, 291–92
  Lou Hawthorne and, 274–75, 279
  lawsuits, 275, 282–83
  and Lee, 276
  and McKinney, 230–236, 250–51, 256–60
Robespierre, Maximilien, 192
Rogers, Roy, 114, 122
Roh, Moo Hyun, 200
Roh, Sung Il, 198, 199
Roles of dogs, 15
Rorvik, David, 239–40
Roslin Institute, 127, 136
  cloned sheep, 22, 126
Ruppy (transgenic beagle clone), 225

Saint-Exupéry, Antoine de, 285
Salamanders, cloning experiments, 130–32
Saltwater agriculture, John Sperling and, 53
San Jose State University, 50
Sapsaree Preservation Association, 223
Sapsarees (Korean dogs), 39, 222–25
Satoh, Yuji, 226
Schatten, Gerald, 127, 177, 197, 198, 200
Schwarzenegger, Arnold, 286
Science, and cloning of animals, 104–5
Science, 196, 200
  and Woo Suk Hwang, 202
Scientific American, and Woo Suk Hwang, 196–97
Sea urchins, cloning of, 128, 129
Seaphire company, 53
Search-and-rescue dogs, 95–96, 102–3
  Trakr clones, 294
Second Addition (goat clone), 105
Second Chance (bull clone), 65–69, 105, 166–73, 289–90
Selective breeding, 136
Seoul National University (SNU), 141–51, 276, 278
  dog cloning, 175–83, 193–95, 225–27, 256
    Booger, 215
    commercial, 208–10
    Sapsaree, 223–25
  and Woo Suk Hwang, 193, 201–2
  and RNL Bio, 189
  and Snuppy patents, 274
Service dogs, 34–35
Seuss, Dr., 104
Shaw, George Bernard, 248
Sheep, cloning of, 135–38
Shelley, Mary, 184
Shields, Scott, 101, 269–70
Shin, Taeyoung, 81, 83, 148, 207, 288
Show dogs, 14
Silver, Lee, 22

Index

Simba (surrogate mother dog), 175, 182
Simon (game), 240–41
Sixth Day, The (movie), 17, 285–86
Snow (cloned Pekingese), 291–92
SNU. *See* Seoul National University
Snuppy (dog clone), 174–77, 180, 182–83, 192–94, 201, 256, 293
Snuppy patents, 274
Snuwolf (wolf clone), 277
Solter, Davor, 134
Somin, Paul, 99–100
Song, Hyon, 141
Sooam Biotech Research Foundation, 203, 206–10, 227–28, 270–72, 276, 278
  and animal welfare, 280
  RNL and, 274, 275
Sophie (Oprah Winfrey's dog), 119
Soros, George, 53
South Korea
  cloning of dogs, 174–83, 219, 277–78, 280, 286–88
  cloning research, 141, 193–94
  dogs as food in, 219–21
  hero dogs, 39
  See also Seoul National University (SNU); Sooam Biotech Research Foundation
Spacecraft, Vorilhon and, 72–74
Speers, d'Armond, 211
Spemann, Hans, 129–32
Sperling, John, 20–22, 26, 44, 47–55, 207, 208, 274
  and animal cloning, 281–82
  and cloned bull, 66–67
  funding of cloning projects, 83, 105
  and Joan Hawthorne, 88–89
  and Lou Hawthorne, 92
  and Texas A&M project, 45, 47, 54
Sperling, Peter, 54
Start Licensing, 208–9, 275, 278
  and Dolly patents, 274
Stem-cell lines, human, cloning of, 176, 179, 180, 196
  problems with, 197–204
Stem cells, human, banking of, 288
Sudworth, John, 234–35
Summum, 123–24
Sunday Telegraph (UK), 72
Surplus of cloned dogs, 281
Symington, James, 11–13, 37–38, 41, 215, 293–94
  and clones of Trakr, 264, 265–66, 268
  and 9/11, 93–94, 102

Tabouleh (cat clone), 152–53
Tahini (Skye Hawthorne's cat), 152
Tai (donor dog), 175, 180–82
Taxidermy, 116, 121–22
Team Trakr Foundation, 294
Television, Second Chance and, 167
Telomeres, 126
Temu, Ron, 123, 124
Texas Agriculture Experiment Station, 46–47
Texas A&M University, 45, 61–62, 290
  cloning of animals, 105–13
    bull, 58, 63–66
    dogs, 80–81, 141, 175
  and Genetic Savings & Clone, 83, 106–10, 111–13
  Shin and, 148
Texas Prison Rodeo, Huntsville, 57
This American Life (TV show), 167, 170
Tigers, cloning of, 146–47, 277
Time magazine
  and cloned dog, 177
  and Woo Suk Hwang, 196
  and hypoallergenic cats, 158
Tippy (collie), 1–4
Titan Protector Ultra (guard dog), 158
To Lift a Nation (sculpture, Watts), 100
Today (TV show), 235, 240
Toppy (clones of Chaser), 225
Totipotency of cells, 129
Tough Guy (mastiff), 29–31, 35, 114–15, 186, 255–56
Tracy (transgenic sheep), 136
Trakr (police dog), 11–13, 36–38, 41–42, 267, 269
  clones of, 263–66, 268–72, 294
  cloning of, 42–43, 215
  and 9/11, 93–95, 102
Trans Ova Genetics, 190
Transgenic animals, 136–37, 150
  beagle clones, 225
  cloning of, 287–88
  pigs, 276
  at Sooam laboratory, 272

Index

Travolta, John, 119
Trigger (Roy Rogers's horse), 122
Trouble (Helmsley's dog), 24
Truatt (Trakr clone), 268
Tuchman, Gary, 98–99
Tumbleweed (Texas longhorn), 58, 61
Turnspit dogs, 216–17
Twain, Mark, 27, 152

UFOland, 74
Uni (clone of Blue Dragon), 224–25
Union organizing, and John Sperling, 53
United Animal Nations, 160
United States
  restrictions on human cloning, 196
  stem-cell research, 289
United States Department of Agriculture (USDA), 159–60
University of Phoenix, 20, 50, 52
  and Biosphere 2, 90
University of Wisconsin, 135
"Upon a Dog Called Fuddle," 214
URAssist, 290
USA Today, 177
Uyeno, Eisaburo, 39

Valiant Ventures, 71
Vaughn, Vince, 60
ViaGen, 164, 187–90, 191
VirtualPetCemetery.org, 119
Vonnegut, Kurt, 155
Vorilhon, Claude (Raël), 70–76, 78, 155–56, 289
  and 9/11, 101

Wall Street Journal, The, 177
War on drugs, John Sperling's views on, 53
Watts, Stan, 124
  To Lift a Nation, 100
Websites. *See* Internet
Weissman, August, 129, 131
West, Michael, 156
Westhusin, Mark, 45–47, 66, 81, 105–7, 170, 173, 178
  and cloning of bull, 62, 63
  and GS&C, 83
  and McKinney, 187–88
  and pet cloning, 112–13
Wharton, Edith, 229
Willadsen, Steen, 134–35
Wilmut, Ian, 35, 46, 125–27, 135–39, 200, 277, 289
  McKinney and, 115
Winfrey, Oprah, 119
Wolfcom Enterprises, 292
Wolfie (Onruang's dog), 292–93
Wolves, cloned, 195, 277
Woolly mammoth, cloning attempt, 127
World Stem Cell Hub, 180, 196
World Trade Center attack (2001), 94–103
  aftermath, Trakr and, 41
W. R. Grace & Co., 135
Wright, Steven, 70

X-inactivation, 108